Corporate Social Responsibility and the State

Jane Lister

Corporate Social Responsibility and the State

International Approaches to Forest Co-Regulation

June 06, 2011

For Peter,

with tremendous gratitude for all your support and guidance.

Jane.

UBCPress · Vancouver · Toronto

21 20 19 18 17 16 15 14 13 12 11   5 4 3 2 1

Printed in Canada on FSC-certified ancient-forest-free paper (100% post-consumer recycled) that is processed chlorine- and acid-free.

Library and Archives Canada Cataloguing in Publication

Lister, Jane
Corporate social responsibility and the state : international approaches to forest co-regulation / Jane Lister.

Includes bibliographical references and index.
ISBN 978-0-7748-2033-2

1. Forest products – Certification. 2. Forest management – Standards. 3. Forest policy. 4. Sustainable forestry. I. Title.

SD387.S69L58 2011   634.9'20218   C2011-902359-8

Canada

UBC Press gratefully acknowledges the financial support for our publishing program of the Government of Canada (through the Canada Book Fund), the Canada Council for the Arts, and the British Columbia Arts Council.

This book has been published with the help of a grant from the Canadian Federation for the Humanities and Social Sciences, through the Aid to Scholarly Publications Program, using funds provided by the Social Sciences and Humanities Research Council of Canada, and with the help of the K.D. Srivastava Fund.

UBC Press
The University of British Columbia
2029 West Mall
Vancouver, BC V6T 1Z2
www.ubcpress.ca

To my loving mother, Marilyn Jane Lister
(1937-2004)

Contents

Illustrations

Tables

Figures

Acknowledgments

This book is a culmination of my twenty years of experience working and researching in the field of corporate social responsibility (CSR) and sustainable resource management. This experience included employment with the Ontario Ministry of Environment, where I worked with industry and environmental organizations on the design and delivery of public/private environmental management programs. I was also a management consultant with the PricewaterhouseCoopers global forest industry group, where I helped companies and governments understand corporate social responsibility, implement environmental management systems, and achieve forest certification. In addition, through this period, I obtained MBA and PhD degrees at the University of British Columbia, delving into the theory behind the practice of CSR. This book therefore offers a first-hand pragmatic and theoretical lens on the topic of CSR governance. I am grateful to the many people who generously shared their insights and helped to shape my ideas along the way.

I am indebted most of all to my academic mentor, Professor Peter Dauvergne, for his encouragement and support of my research and for continually raising the bar on my efforts, understanding, and publishing aspirations. I am also privileged to have had ongoing helpful direction from Professor Peter Nemetz at the Sauder School of Business and from Linda Coady, distinguished fellow at the Liu Institute for Global Issues at UBC. In addition, I was fortunate to receive very thorough and thoughtful feedback on my work from Professors John Innes, Gary Bull, and Kernaghan Webb.

I owe a great debt of gratitude to the faculty, staff, and students at the Institute for Resources, Environment and Sustainability (IRES), who provided a stimulating interdisciplinary home for my research. In particular, I am grateful to Les Lavkulich, the founding director, and Gunilla Öberg, the current director. I also appreciate the helpful guidance provided by a number of professors across the UBC campus, including, in particular, Kathy Harrison,

George Hoberg, Alan Jacobs, and Rob Kozak. I am grateful to several scholars and practitioners from outside of UBC for their insights and advice at critical stages of my research, including Michael Howlett (Simon Fraser University); Lars Gulbrandsen (Fridtjof Nansen Institute); Connie McDermott, Graeme Auld, and Ben Cashore (Yale University); Karin Lindahl (Swedish University of Agricultural Sciences); Tage Klingberg (University College of Gävle); Bill Cafferata (retired Chief Forester, MacMillan Bloedel); Bruce Eaket (PricewaterhouseCoopers); and John Braithwaite (Australian National University).

I am thankful for the opportunities I had to participate and receive feedback at several academic workshops, including initial guidance from Atle Middtun (Bedriftsøkonomisk Institut [BI], Norwegian School of Management) and Ed Freeman (University of Virginia) at the CSR PhD Workshop at the Copenhagen School of Business (CBS) sponsored by the European Academy of Business in Society (EABIS) in October 2004; exploration of the emerging private environmental governance research agenda with Robert Falkner, Andy Gouldson, and Philip Pattberg and their graduate students at the London School of Economics Role of Private Actors in Global Politics Workshop in November 2005; and valuable feedback on my research methodology at the Dartmouth College Workshop on Industry Self-Regulation in February 2006 from, in particular, Andrew King (Dartmouth), Michael Toffel (Harvard University), Andy Hoffman (University of Michigan), Aseem Prakash (University of Washington), and Cary Coglianese (University of Pennsylvania).

Several awards and grants supported my research, including a UBC Faculty of Arts US Studies Weyerhaeuser Foundation Research Grant, an Environment Canada Applied Environmental Economics and Policy Research Scholarship, a Social Sciences and Humanities Research Council of Canada (SSHRC) Doctoral Fellowship Award, and generous financial support provided by Peter Dauvergne under his SSHRC-sponsored project on the Global Environmental Politics of Corporate Social Responsibility.

The research would not have been possible without the participation of the many interviewees across Canada, the United States, and Sweden who were incredibly generous with their time and insights. In particular, I am indebted to J.P. Kiekins (ForestWatch), Jeff Serveau (Industry Canada), and Rick Fox (US Forest Service) for directly facilitating my research by inviting and supporting my attendance and participation at their forest conference and delegation meetings.

This book has been published with the help of a grant from the Canadian Federation for the Humanities and Social Sciences, through the Aid to Scholarly Publications Program, using funds provided by the Social Sciences and Humanities Research Council of Canada. I am very grateful to the two anonymous reviewers for their very thorough and thoughtful feedback on the manuscript and to Anna Eberhard Friedlander, production editor, and Randy Schmidt, senior editor, at UBC Press for their supportive guidance through the publishing process.

Finally, I am most thankful to my family and friends for anchoring me and reminding me of my tremendous good fortune.

Abbreviations

AF&PA	American Forest and Paper Association
AFS	Australian Forestry Standard
ATFS	American Tree Farm System
ANSI	American National Standards Institute
BCTS	British Columbia Timber Sales
BLM	Bureau of Land Management (US)
BMP	best management practice
CCFM	Canadian Council of Forest Ministers
CFSA	Crown Forest Sustainability Act (Ontario)
CLFA	Crown Lands and Forest Act (New Brunswick)
CPET	Central Point of Expertise on Timber Procurement
CORE	Commission on Resources and Environment (British Columbia)
CPPA	Canadian Pulp and Paper Association
CSA	Canadian Standards Association
CSR	corporate social responsibility
DCR	Department of Conservation and Recreation (Massachusetts)
DNR	Department of Natural Resources
EAA	Environmental Assessment Act (Ontario)
EMS	environmental management system
ENGO	environmental nongovernmental organization
EPA	Environmental Protection Agency
EU	European Union
FLEGT	Forest Law Enforcement, Governance and Trade
FAO	Food and Agriculture Organization of the United Nations
FPAC	Forest Products Association of Canada
FPB	Forest Practices Board (British Columbia)

FRPA	Forest and Range Practices Act (British Columbia)
FSC	Forest Stewardship Council
GRI	Global Reporting Initiative
HCP	Habitat Conservation Plan
HCVF	high conservation value forest
ISEAL	International Social and Environmental Accreditation and Labelling Alliance
ISO	International Organization for Standardization
ITTO	International Tropical Timber Organization
LEED	Leadership in Energy and Environmental Design
LRMP	Land and Resource Management Plan (British Columbia)
MCPFE	Ministerial Conference on the Protection of Forests in Europe
MFL	Managed Forest Law (Wisconsin)
MNC	multinational corporation
MRNF	Ministère des Ressources naturelles et de la Faune (Quebec)
MTCC	Malaysian Timber Certification Council
NBF	National Board of Forestry (Sweden)
NGO	nongovernmental organization
NIPF	non-industrial private forest
NRDC	Natural Resources Defense Council
NSF-ISR	NSF-International Strategic Registrations
NSMD	non-state market-driven
OECD	Organisation for Economic Co-operation and Development
OMNR	Ontario Ministry of Natural Resources
OSP	Ontario Stewardship Program
PAS	Protected Areas Strategy (British Columbia)
PEFC	Programme for the Endorsement of Forest Certification
QFIC	Quebec Forest Industry Council
RAN	Rainforest Action Network
REIT	real estate investment trust
SBFEP	Small Business Forest Enterprise Program (British Columbia)
SCC	Standards Council of Canada
SCS	Scientific Certification Systems
SEPA	Swedish Environmental Protection Agency
SFA	Swedish Forest Agency
SFB	Sustainable Forestry Board
SFI	Sustainable Forestry Initiative

SFL	Sustainable Forest Licence (Ontario)
SFM	sustainable forest management
SIC	SFI implementation committee
SSNC	Swedish Society for Nature Conservation
SWEDAC	Swedish Board for Accreditation and Conformity Assessment
TFL	tree farm licence
TIMO	timber investment management organization
TSFMA	Timber Supply Forest Management Agreements (Quebec)
UNCED	United Nations Conference on Environment and Development
UNECE	United Nations Economic Commission for Europe
UNFF	United Nations Forum on Forests
USFS	United States Forest Service
WBCSD	World Business Council for Sustainable Development
WCED	World Commission on Environment and Development
WTO	World Trade Organization
WWF	World Wide Fund for Nature/World Wildlife Fund

Corporate Social Responsibility and the State

1
Introduction

Over the past fifteen years, private environmental codes and transnational corporate social responsibility (CSR) standards have proliferated. Led by industry and/or nongovernmental organizations (NGOs), these standards now address sustainability issues in a wide range of sectors across the globe – from forestry, mining, oil and gas, fisheries, agriculture, finance, and chemicals to apparel, coffee, jewellery, and tourism. Governments, corporations, and NGOs have been enthusiastic about CSR, with many groups heralding these voluntary multi-stakeholder efforts as the path to sustainable development.

The CSR opportunity is enticing to all stakeholders. When a corporation voluntarily takes on greater responsibility for achieving societal goals, its long-term value can increase, negative environmental impacts ideally are reduced, and the regulatory costs to governments are ultimately lessened. It can be a win/win/win scenario. As CSR participation unfolds in a patchy, uneven pattern, however, and as environmental and social conditions worsen in vulnerable areas across the planet, skepticism about CSR standards is growing. The sense is that, on their own, they are falling short. Attention is shifting back to governments to "scale up" CSR efforts. Some states have heeded the call, whereas others remain on the sidelines. The role of the public sector is unclear and is a point of global debate. Should governments ignore, facilitate, compete with, or perhaps even mandate these voluntary private standards?

On the one hand, if they enable CSR, governments could be perceived as handing over the policy reins – effectively turning the fox loose in the henhouse in trusting the market with the public good. On the other hand, if they ignore or merely observe CSR, governments may lose the opportunity to leverage private resources for public benefit as well as to reward corporate virtue. The implications of state engagement with CSR are largely unexplored. This book addresses this knowledge gap.

The following seven chapters assess the public sector role in CSR by evaluating government response to a well-established CSR standard, forest certification. The focus is on forest certification not just because it is a highly developed example of CSR but also because the pattern of certification adoption and its policy classification are puzzling: if certification was intended to fill a governance gap in lesser-developed tropical regions, why has 90 percent of adoption taken place in highly regulated, northern developed countries? And why have governance scholars and policy makers labelled certification a non-state, market-driven mechanism when the standards incorporate public forest laws and governments are directly engaging with the certification process?

This is the comprehensive story of forest certification governance that goes beyond a market-based narrative. Although governments have generally assumed a position of non-interference in forest certification, they are in fact responding to certification through a range of direct co-regulatory approaches – endorsing, enrolling, and even mandating the private governance mechanism as an additional policy tool alongside traditional forest regulation. Furthermore, in the highly regulated, developed countries where forest certification co-regulation is occurring, the contest between overlapping established public laws and new private forest rules is encouraging ongoing adaptive improvements in forest management policy and practices. In other words, forest certification is aiding governments to better manage their commercial forestry sector and forest resources.

There is optimism regarding the emerging benefits of co-regulation. Private forest owners are taking on greater responsibilities for forest sustainability. Public policy makers are learning from the ongoing contest between public and private rules, and from the increasing level of collaboration. Nevertheless, the story of forest certification governance is also a cautionary tale that highlights the limits of CSR. Although it is perhaps tempting for a government to employ forest certification as a replacement for often financially costly forest regulations and programs (such as planning, inventories, and audits), the cases in this book reveal how forest certification is *not* a substitute for the efforts of a public forest agency. Rather, success in forest certification hinges on state capacity and government engagement.

The remainder of this chapter is divided into three sections. The first section introduces the topics of CSR and forest certification, defines the concept of CSR co-regulation, and outlines the central arguments regarding certification co-regulation. The next section explains the objectives and parameters of the research, specifically the reasons for selecting Canada, the United States, and Sweden as the case study examples. The chapter concludes with a brief overview of the book's structure.

Corporate Social Responsibility

Corporate social responsibility is fundamentally about the role of business in society and the balancing of public and private responsibility. To what extent does a company have a responsibility to go beyond the law to meet societal expectations? Is the corporate mandate solely to deliver a financial profit to shareholders, or do businesses also have a responsibility to create value for society? Are the two goals mutually exclusive? There is a long history of debate over these normative questions, with shifting emphases and fluctuating levels of societal concern.[1]

Since the United Nations Conference on Environment and Development (UNCED) in 1992, there has been a resurgence of interest in CSR, with attention directed in particular towards increasing the accountability and responsibility of multinational corporations (MNCs) in addressing environmental and social equity issues and contributing to global sustainability solutions. Groups across all sectors have reacted. The UN spearheaded the United Nations Global Compact to encourage multinational companies to voluntarily commit to the adoption of global CSR principles. The International Organization for Standardization (ISO) established the ISO 14000 set of international environmental management standards to help companies mitigate their environmental risks. NGOs initiated the Global Reporting Initiative (GRI) to provide a template for more standardized corporate sustainability reporting. Financial institutions introduced sustainability funds and socially responsible investment rating systems such as the FTSE4good and the Dow Jones Sustainability Group Index to identify and send stronger market signals about responsible companies. And industries from resource extraction to retail have developed and adopted multi-stakeholder CSR standards in their respective sectors.

Led in many cases by environmental advocacy organizations and companies in partnership, many of the CSR standards include certification programs that aim to encourage sustainable business practices by linking responsible producers and consumers through market supply chains. Independent auditors verify and certify a company's stewardship practices based on a checklist of environmental and social requirements spelled out in the private certification standards. Certified producers can then carry an eco-label on their products, providing them with opportunities to access growing eco-consumer markets.

Going beyond legal requirements, these standards have become prevalent across an increasing number of industry sectors. For example, the Marine Stewardship Council certifies and labels sustainably harvested seafood from wild fisheries. Coffee, cocoa, and tea producers seek sustainability labels of approval under international fair trade and organic certification bodies, including the private UTZ and Rainforest Alliance programs. RugMark International certifies carpet producers for responsible labour practices, including the elimination of child labour. The Global Sustainable Tourism Council oversees standards for certifying environmental and socially responsible tourist operations. The multi-stakeholder Initiative for Responsible Mining Assurance independently verifies compliance with environmental, human rights, and social standards for mining operations. And roundtables in many sectors, such as biofuels, palm oil, and soy, are defining requirements and developing standards and eco-labels to verify sustainable commodity production. Business practices around the world are now being shaped by these private environmental governance mechanisms.

CSR standards constitute a distinct form of environmental governance, with rule-making capacity that goes beyond traditional industry self-regulatory codes of practice (such as those in the financial, medical, and media professions). For example, many include democratic multi-stakeholder decision-making bodies that operate by consensus under the terms of written constitutions; innovative sustainability requirements that go beyond the law and that are regularly revised through public consultation; third-party audits conducted by independent professional auditors to ensure ongoing compliance; and an eco-label to create market incentive. As this book explores in the case of forest certification, these standards present an unprecedented co-regulatory governance challenge and opportunity.

Forest Certification and CSR Co-Regulation

Forest certification is a multi-stakeholder, voluntary CSR initiative that encourages sustainable forest management (SFM)[2] by leveraging market supply chains to link customer demand for certified forest products with responsible producer supply. To achieve certification, a forest operation undergoes an independent third-party audit to verify that the forest is managed in accordance with a checklist of ecological, economic, and social sustainable forest management principles spelled out in a forest certification standard. The certification standard is developed and revised by a multi-stakeholder decision-making body made up of a range of members representing different forest values and interests. Once an operator is certified, its forest and paper products can be awarded an eco-label by means of a chain-of-custody certification audit that traces the certified fibre down through the product supply chain from the forest to the end customer. Ultimately, forest certification aims to provide producers with a market incentive to adopt improved sustainable forestry practices.

Since the first forest management certifications in the 1990s (under the current third-party audited standards), many private companies and family forest owners as well as government public landowners have signed on, and certification continues to increase steadily. As shown in Figure 1.1, over the

Figure 1.1

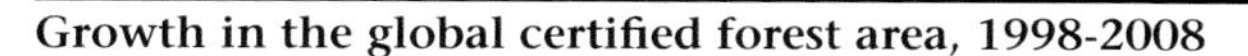

Growth in the global certified forest area, 1998-2008

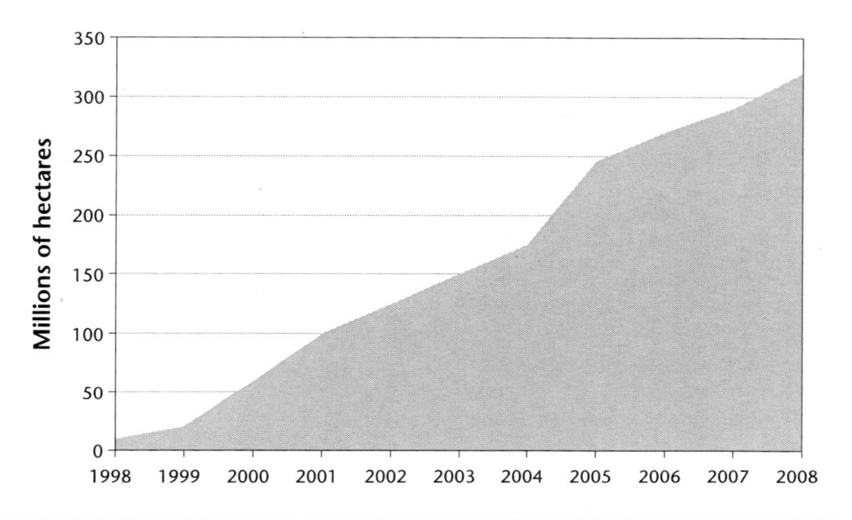

Source: UNECE/FAO 2008, 109.

past decade the area of global certified forest expanded from approximately 10 million hectares in 1998 to 320 million hectares in 2008. Although the rate of adoption has recently slowed, the total area of certified forest around the globe continues to grow. Consequently, forest certification remains an important concern for the global forest industry and timber-producing nations.

Environmental nongovernmental organizations (ENGOs) initiated forest certification in the late 1980s and early 1990s in order to curtail deforestation and forest degradation, particularly in the lesser-developed tropical regions of the world that lacked sufficient forest regulation. Instead of addressing a capacity gap in developing countries, however, certification systems have been adopted as an overlapping forest governance mechanism in *developed* countries. For example, whereas over 50 percent of Western European forests and over 33 percent of North American forests are certified, less than one-tenth of 1 percent of forests in Africa and Asia are certified, and only 1.6 percent of Brazilian forest is certified (of which the majority is plantation forest) (UNECE/FAO 2008, 107). Factors explaining the low level of certification adoption in the Global South include high financial costs, complex property rights, inadequate regulatory capacity, and weak market signals (Cashore et al. 2006). Although certification has begun to slowly increase in lesser-developed countries, so far it has not been an effective governance mechanism to combat tropical deforestation. Instead, it is succeeding in promoting continual improvement in sustainable forestry practices in already highly regulated northern boreal and temperate regions of North America and Western Europe.

Rather than dismissing the pattern of lagging certification adoption in the South as a global regulatory disappointment or failure, this research has seized upon a window of opportunity to investigate the significance of public sector capacity to CSR. The inquiry is spurred by several underlying questions. Given that forest laws are already well established in the industrialized countries where certification is occurring, what regulatory purpose is certification actually serving? If certification is gaining a regulatory foothold, does this imply that the state has retreated? How are governments responding to certification, and why is there a variation in response? To what extent is the dynamic between public and private forest rule-making authority competitive versus cooperative? And finally, is co-regulation making a difference – are there positive forest management outcomes?

Traditional "statists" have interpreted the emergence of private authority as a retreat of the state, or governance without government. Political authority is assumed to be a zero-sum contest.[3] Contrary to this theoretical perspective, however, the empirical reality of forest certification governance demonstrates the coexistence of public and private authority. Governments are actively engaging in and even mandating certification. This book argues, therefore, that private environmental authority in the case of forest certification does not constitute a retreat of the state but rather a shift in the role of government towards greater multicentric governance. Until recently, political scholars have largely ignored this governance transformation. Governments *are* engaging in CSR private standard setting, and CSR is serving a policy role, but we have very little empirical or theoretical understanding of these newly forming "post-sovereign" co-regulatory governance systems.

In the absence of a theory of CSR governance, I introduce the concept of *CSR co-regulation* in order to provide an analytical lens through which to identify and assess the emerging public/private shared governance arrangements. CSR co-regulation refers to state engagement with CSR standards alongside public regulation so as to leverage private resources and initiative.[4] Beyond introducing the concept, I explain the process and challenges of CSR co-regulation, and present three new analytical tools to support and guide the case study evaluations of co-regulatory governance. The first is a governance typology that classifies and highlights the unique aspects of CSR standards such as forest certification among the array of traditional regulatory and emerging public/private cooperative policy instruments (see Figure 2.1). The second is a matrix for illustrating the overlapping public/private boundaries of CSR standards with traditional and emerging forms of regulation (see Figure 2.3). The final tool is a framework for mapping government response to CSR along a spectrum of engagement at the various stages of the policy cycle (rule-making, implementation, and enforcement) (see Figure 2.4). All three tools help to illuminate the concept of CSR co-regulation and reveal how it is occurring in practice.

Governance scholars have attached many labels to certification in order to emphasize its private regulatory capability. Terms include "civil regulation," "private hard law" (versus non-prescriptive *soft* law), and "non-state market-driven (NSMD) governance." NSMD has gained acceptance in the certification literature. Under the NSMD theory, CSR standards such as forest certification are considered purely private mechanisms, establishing private rules independent of state authority (Cashore 2002; Cashore, Auld, and Newsom 2004). As even the theory acknowledges, however, certification systems rely on a baseline legal framework, require regulatory compliance, and also incorporate formal international state-based sustainability principles. What's more, government authorities are overseeing, facilitating, legitimating, and, in some cases, even enforcing certification. Certification is neither purely private nor purely market-driven. It is both. This book therefore challenges the accepted NSMD theory, arguing that it is a partial classification. A transnational multi-stakeholder standard such as forest certification *is* unique with respect to its non-delegated private authority, but it also overlaps with public authority and, most importantly, achieves acceptance and success through state capacity and government engagement. In other words, CSR standards such as forest certification are more accurately classified as co-regulatory governance mechanisms.

By evaluating the role of government in forest certification in the leading global certified nations (Canada, the United States, and Sweden), the cases included in this book demonstrate that although certification has weaknesses as a stand-alone forest policy instrument (for example, it does not address overall forest health and does not necessarily fully align with broader government forest objectives), it can provide a supplementary regulatory resource. This includes potential contributions to the three key areas of governance: polity (decision-making forum), politics (decision-making process), and policy (decisions). Specifically, certification can expand the political arena, facilitate greater multi-stakeholder deliberation, and encourage more innovative forest rules. By strategically combining the dynamism and innovation of private certification standards with the stability and democratic accountability of traditional state-led regulatory approaches, certification co-regulation can establish a more flexible range of options for policy makers, which can in turn encourage continual regulatory improvements. Overall, this book argues that in developed countries with high state capacity, certification co-regulation can constitute a progressive step towards more responsive and adaptive rule making, and hence more effective collective sustainability solutions.

Figure 1.2

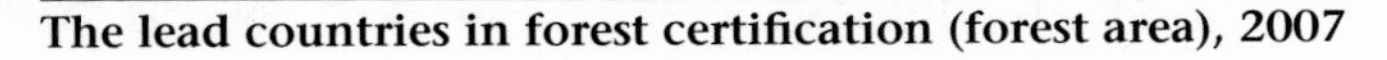

The lead countries in forest certification (forest area), 2007

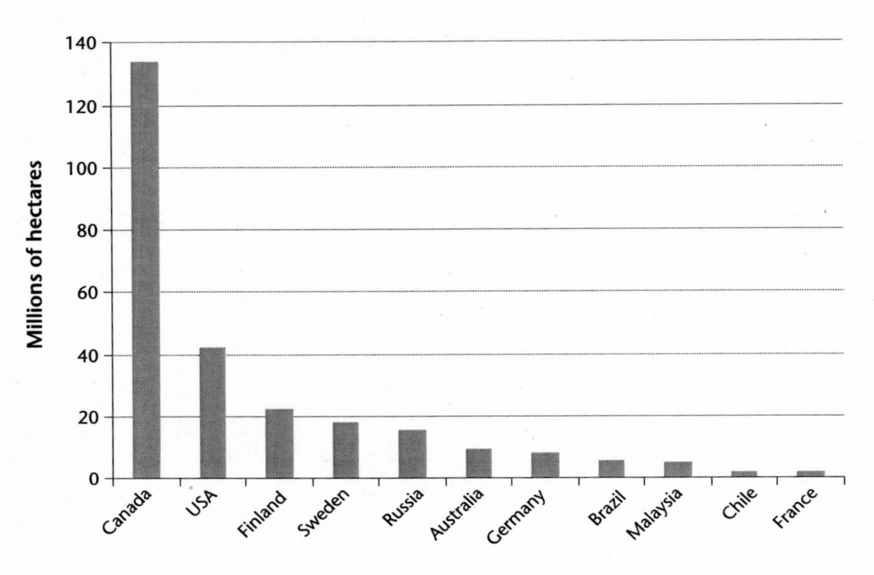

Source: UNECE/FAO 2008, 111-12.

The Leading Certified Nations

Canada, the United States, and Sweden are major global forest product producers and exporters. They have also been global certification leaders in terms of their forest certification development and implementation efforts (see Figure 1.2). For example, Canada and the US account for the majority of the world's certified forest, and have also developed the world's leading national certification programs in terms of total certified area (e.g., the Sustainable Forestry Initiative [SFI] in the US, and the CAN/CSA-Z809 standard in Canada; see Appendix B). Sweden has also been a leading country in terms of certification adoption (achieving certification of over 60 percent of its forests), and has served as a flagship for the Forest Stewardship Council (FSC) certification program (initiating and adopting the first national FSC standard in 1998). In addition, these three countries have long histories of well-established yet varying forest regimes (Table 1.1).

Canada, the US, and Sweden are therefore logical cases to choose in conducting a certification governance analysis. Global timber production is an important selection criterion because certification achieves leverage through global supply chains. Certification leadership is essential, as certification needs to have gained a sufficient foothold in the region in order for co-regulation to be studied. Finally, varying forest regimes within the sample

Table 1.1

Case study forest regimes

Country	Forest regime
Canada	Highly regulated at the provincial level, with majority public land
United States	Variable regulatory approaches at the state level, with majority private land
Sweden	Highly regulated at the national level, with majority private land

of cases provide an opportunity to examine the institutional influence of baseline regulatory structures on the co-regulatory policy dynamic.

An apparent omission from the research sample is Finland. Although Finland meets the case selection criteria, this large Nordic timber producer is not addressed, as my intent here is to focus on the "hard cases" with the greatest public/private tension (i.e., rule-making contest). Very early on, in April 1996, the Finnish government took a direct leadership role in initiating the development of a national certification standard based on the country's national forest program.[5] Thus, rather than tension in deliberations over public and private forest rules, what occurred in Finland was that over 95 percent of forestland was certified to the Finnish Forest Certification standard within two years of its approval in 1998. The forest certification co-regulatory dynamic in Canada, the US, and Sweden has been much more complex.

Another apparent omission is the southern tropical countries. Although it might seem intuitively obvious to focus an investigation into forest governance on places where the worst global forest degradation and deforestation problems are occurring (such as the tropical forests), the lack of consistently strong forest institutions and certification adoption in these developing regions limits the potential for a co-regulation analysis. As the aim of this research is to understand the dynamic of interacting public and private authorities, it is critical that the cases constitute jurisdictions with both high public and high private governance capacity.[6] High regulatory function and capability provide the context for the greatest potential public/private political tensions, and this enables an evaluation of co-regulation challenges, benefits, and optimal arrangements.

As shown in Figure 1.3, ensuring both high public and high private capacity places this research firmly within *developed* rather than *developing* countries.[7] It is important to note, however, that this book nevertheless has a bearing on forest governance decisions in the developing countries, where the most rapid global forest loss is occurring. This is particularly true in terms of the book's insights into the importance of public capacity, and the

Figure 1.3

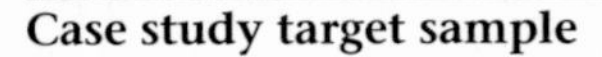

Case study target sample

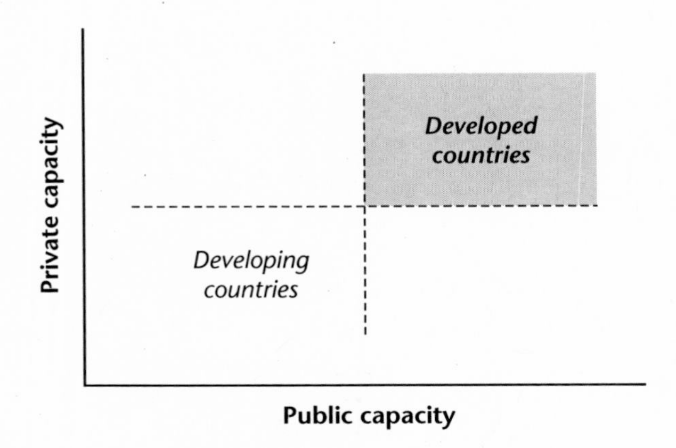

cautionary warnings about how certification falls short as a substitute for well-designed and delivered government forest programs and regulation.

Because forest regulatory responsibility resides at the subnational level within Canada and the US, these cases include analyses of provincial and state governments. Within Canada, these include British Columbia on the west coast, Ontario and Quebec in central Canada, and New Brunswick on the east coast. Within the US, these include the twelve state governments across the country that had certified their state-owned forestlands as of 2008 (Massachusetts, Pennsylvania, Maryland, New York, North Carolina, Michigan, Minnesota, Wisconsin, Maine, Tennessee, Washington, and Indiana).

The four Canadian provinces were selected because they are the top forest product producing regions in Canada, and present a range of government certification responses and industry expectations of government's role (for example, two of the four provinces have mandated certification).[8] All twelve US certified states are included as this permits an assessment of the range, interaction, and implications of the various factors influencing each state's similar co-regulatory response. And Sweden is addressed as it provides a clear example of government's role in facilitating the competitive and cooperative forest certification co-regulatory dynamic.

Objectives and Parameters

As explained, this book presents three in-depth case studies that explore Canadian, US, and Swedish government responses to certification. It provides a historical snapshot of the period 1993 to 2008. Five objectives shaped the case study investigations:

- assess the emergence, evolution, and adoption of the leading certification programs
- identify and compare government responses to forest certification
- compare the rationale and drivers of government certification engagement
- analyze the dynamic of certification/forest policy interaction
- evaluate the forest governance implications of certification co-regulation.

The arguments and findings here are supported by evidence I gathered over a three-year period (2004-07) through interviews with over 120 key forest governance stakeholders and experts across Canada, the US, and Sweden (see Appendix A). The three central questions that I asked of the interviewees were:

- How has government responded to certification?
- Why has government adopted its particular certification co-regulatory approach?
- How has certification co-regulation affected forest governance?

This book offers insights into CSR co-regulation governance within specific research parameters. Six key areas define the research scope.

First, although the central aim of this book is to understand the broader question of the public sector role in CSR co-regulation, the research is concentrated on the specific case of government response to forest certification. As already noted, this is logical as forest certification is a well-developed and established example of a CSR governance standard. As private environmental governance standards gain institutional strength in other industry sectors, there will be future opportunities to apply the analytical framework presented here in order to compare government co-regulatory responses to several different CSR standards within and across political jurisdictions.

Second, unlike most forest certification governance studies, which address only the Forest Stewardship Council, this analysis concerns both the Programme for the Endorsement of Forest Certification (PEFC) and FSC systems. As of 2008, PEFC and FSC international certification programs accounted for 68 percent and 32 percent of the total global certified forest area, respectively. Although strong philosophical differences remain and debate continues over the various strengths and weaknesses of the PEFC versus FSC standards, over the past decade the two systems have been generally converging in terms of their multi-stakeholder participation, independent audit requirements, and SFM indicator content. Governments have taken officially neutral positions regarding support for one system over another.

Environmental organizations continue to recognize only the FSC. Major retailers tend to accept either standard as legitimate and credible, while stating a preference for FSC-certified products if they are available. Producers have adopted both PEFC and FSC systems in all of the case study regions, with more and more examples of "dual certification." Consequently, unless otherwise specified, throughout this book the terms "certification" and "certification co-regulation" are inclusive and refer to both PEFC and FSC programs.

Third, in terms of the level of the analysis, the cases focus on the level of the bureaucracy (the lead forest departments and agencies within each jurisdiction) where co-regulatory policy development and implementation occurs. Broader speculative political questions regarding the degree of influence of the type of state and form of government (e.g., presidential versus parliamentary; federal versus unitary); the role of party politics (e.g., left- versus right-of-centre); or the comparative contribution of executive, legislative, and judicial actors in certification co-regulation are not systematically evaluated.[9] As all of the case study jurisdictions are democratic, however, and the bureaucracy is an agent of the elected government, concentrating at the level of the forest agency not only provides a means of evaluating certification co-regulation policy formulation and delivery but also permits an understanding of the influence of elected officials and internal administrative politics.

Fourth, although the cases provide an essential research parameter, the book is also bounded by the research questions. As the aim is to evaluate the interaction of public and private rule-making systems, the case evaluations focus on how and why certification co-regulation is occurring and the governance outcomes. The analysis does not include the on-the-ground effectiveness of certification co-regulation (i.e., the difference that a shared governance approach is making towards resolving specific forest problems such as deforestation, illegal logging, forest conversion, biodiversity preservation, endangered species, carbon storage, or Aboriginal rights). The effectiveness with respect to the positive *forestry management* governance outcomes is addressed, but the evaluation excludes actual *forest* outcomes.[10] Few studies have yet to tackle the question of certification "problem-solving" effectiveness, primarily because it has been too early to assess the impacts. In addition, measuring impact is difficult because forest certification requirements focus on improving management processes and site-level forestry practices rather than achieving specific landscape-level forest conditions. Isolating and measuring the on-the-ground forest effects attributable to certification versus other factors is inherently complex and uncertain. Given the lack of empirical data, this aspect of certification co-regulation is not included in the case study evaluations. As discussed in the conclusion, however, the effectiveness of certification in changing forest conditions is an important area for future investigation.

Fifth, although the cases are bounded by the same research approach and questions (such as how and why governments engaged in certification and what the implications were), each case presents a slightly different co-regulation puzzle according to local forest regime conditions. The focus in each case is therefore slightly different. This enhances the contextual details but also limits the direct comparison between the cases. For example, in the Canadian case, the compelling question is why, across *similar* forest regulatory regimes, did provincial governments respond differently to certification? In the US, the situation is the opposite. Why did *different* state forest regulatory regimes respond similarly to certification? And in the Swedish case, given the results-based "frame law" policy environment that enabled certification development and adoption, how did certification and public policy interact, and did the Swedish forest authorities retreat and hand off policy responsibility? Although limiting comparability between cases, this slight variance in focus facilitates an important progression. The cases evolve in their depth and focus from a broad examination of the range of government certification roles (Canada) to a concentrated study of the governance implications of a specific co-regulatory approach (US), to an in-depth investigation of the certification co-regulation policy dynamic (Sweden). Consequently, the key analysis occurs within each chapter rather than in a separate comparative evaluation of the cases at the end of the book. A synthesis of the case study results is presented in the conclusion.

Lastly, an important research parameter concerns the governance target. Both the Canadian and US cases concentrate on the governance implications of certification co-regulation on public land. In Canada, this is appropriate because over 90 percent of forestland is publicly owned. In the US, although the majority of forestland is privately owned, it is logical to focus on state government adoption of certification on state-owned, public forestland as state lands account for a surprisingly disproportionate percentage of the total certified forest area across the country. The analysis does not include US private non-industrial forestland, as less than 1 percent of family forest owners have certified their forestland and government certification incentives had only just begun to develop during the study period. As well, the US case focuses on state forests and not the national forests, as the US Forest Service position during the study period was to assess rather than implement certification.

In summary, this book constitutes a piece of a much larger emerging area of investigation concerning CSR co-regulation, and therefore necessarily has distinct parameters. In outlining the focus and boundaries of the research (the cases selected, the questions examined, and the research approach), it is hoped that these parameters will serve as a guide to the extent to which the study results can be generalized to other cases, as well as highlight opportunities for future research.

Structure

The seven chapters of this book develop the central argument that CSR co-regulatory arrangements are emerging whereby governments are engaging in forest certification, integrating private authority within their policy mix to enhance forest governance. The purpose of this first chapter has been to introduce the topic, present the main arguments, and review the research objectives and scope. Chapters 2 and 3 provide the background and the theoretical context for the three case study evaluations. Specifically, Chapter 2 explains the emergence of CSR, defines the concept of CSR co-regulation, and presents a typology as well as a mapping tool for evaluating the shifting regulatory role of the state with regard to the various new modes of co-regulatory governance. Chapter 3 follows a similar progression, but with respect to the particular CSR example of forest certification. It begins by explaining the emergence of forest certification and evaluating its unique classification as a non-delegated private governance mechanism. The chapter then introduces the specific case of certification co-regulation and, in particular, applies the co-regulatory matrix introduced in Chapter 2 to assess the range of regulatory instruments within a co-regulatory forest governance system.

Chapters 4, 5, and 6 comprise the three empirical case studies – Canada, the United States, and Sweden. Each case study has a similar structure. Each evaluation begins with background on the local forest regime and an overview of forest certification development and adoption within the respective jurisdictions. The role of government in forest certification is then assessed in terms of the approach, drivers, and governance implications of certification co-regulation within each jurisdiction. Chapter 7 presents a synthesis of the case findings, an evaluation of the limits and potential of CSR co-regulation, operational recommendations for policy makers on achieving optimal CSR co-regulation, and suggestions for future research.

2
Co-Regulating Corporate Social Responsibility

Not only are corporate social responsibility (CSR) standards becoming increasingly prevalent but many are also gaining unprecedented private rule-making authority and governance function, essentially mimicking the policy role of public institutions. Transnational CSR standards are serving as a global environmental governance mechanism to supplement international laws and agreements, as well as behaving as domestic private regulations alongside established national laws and strong public institutions. This raises an interesting puzzle. How is CSR private authority interacting with state authority in domestic political environments with high public capacity? Are the public and private rule-making systems competing or cooperating? What is the policy role of CSR private standards, and what is the role of government in CSR private rule making in these jurisdictions?

This chapter presents three main arguments. First, CSR multi-stakeholder standards constitute not only a distinct mode of governance but also a new self-regulatory policy instrument. Second, with the emergence of private authority, governments are not in retreat but rather are transforming their role from policy delivery and delegation to also enabling private regulations alongside traditional regulation within co-regulatory governance systems. And third, there is a spectrum of interventions by which governments can co-regulate CSR so as to leverage private initiative and supplement governance capacity.

The chapter also introduces three new analytical tools: (1) a typology to distinguish private environmental governance from the traditional hierarchical and self-regulatory modes of governance; (2) a matrix that illustrates how a co-regulatory governance system includes a mix of private and public policy instruments; and (3) a regulatory scale to identify and position government response to CSR along a spectrum of indirect to direct engagement at the various stages of policy development, implementation, and enforcement. These tools provide an analytical lens for viewing CSR co-regulation as well as a theoretical framework to guide the empirical case study investigations in Chapters 4, 5, and 6. To begin, we turn to the definition and emergence of CSR standards and private governance authority.

CSR and the Emergence of Private Authority

Corporate social responsibility is about the role of business in society, and societal expectations that companies will "do good" by contributing skills, power, and resources to meet global sustainability goals. Nongovernmental organizations (NGOs), governments, and industries around the globe have promoted CSR as a progressive, self-regulatory approach to achieving sustainable development. For example, the European Commission designated 2005 as the year of corporate social responsibility. While there is no single accepted definition of CSR, fundamentally it concerns the voluntary choice of companies to integrate societal concerns (in addition to shareholder interests) into their business operations. Leading companies demonstrate CSR and corporate citizenship by going beyond legal requirements to meet stakeholder expectations regarding a triple bottom line of economic, social, and environmental objectives.[1] Specifically, exemplary corporations undertake *firm-level* sustainability management programs such as environmental auditing and reporting, life cycle assessments, stakeholder consultation, supply chain greening, socially responsible investing, and sustainability reporting. As well, many firms that embrace CSR cooperate in the development and implementation of *industry-level* CSR codes and standards. These are broader accountability and transparency initiatives that encompass general CSR principles as well as firm-level management programs.

CSR is not necessarily formally delegated or enforced by the state, and compliance with the voluntary private standards is achieved through audits, public reporting, oversight by the standards boards, and "naming and shaming" of non-cooperators (free-riders). Corporations are motivated to adopt voluntary CSR initiatives by many factors, including a desire to avoid regulation, reduce risk, and manage corporate reputation in the face of increased public pressures and advocacy group lobbying efforts (Hoffman 2001; Lyon and Maxwell 2004; Prakash 2001). Many companies have also sought to realize the potential win/win "sustainable development" and "ecological modernization" opportunities and advantages of combining economic growth with environmental and social considerations, and green technology innovation as promoted by academics, governments, business associations, and NGOs.[2]

Broadly speaking, a convergence of social, political, and economic factors in both the domestic and global arenas have contributed to the recent emergence of CSR initiatives. At the domestic level, during the 1980s and 1990s, governments in industrialized countries implemented public sector reforms that included performance-driven business management practices within government, outsourcing, and adopting greater self-regulatory policy approaches. The aim was to streamline government and achieve greater efficiencies in public administration and policy delivery.[3]

Running in parallel in the global arena, concerns were building regarding the increased power of multinational corporations (MNCs) and the growing prominence of global human rights, labour, and environmental issues. Problems associated with rapid economic globalization were deemed to be outpacing the governance capacity of state governments and international mechanisms to achieve timely, democratic, and effective outcomes. In reaction, transnational advocacy groups formed to address the global governance gap.[4] Overall during this period, global civil society organizations as well as domestic-level advocacy groups called on corporations to assume increased environmental and social responsibilities.

Corporate response to the global and domestic pressures for greater self-regulatory CSR efforts was mobilized within the United Nations' World Commission on Environment and Development (WCED; established in 1987) and at the United Nations Conference on Environment and Development (UNCED) in 1992. In both forums, governments introduced and promoted the win/win possibilities of sustainable development solutions as a means to help close the global environmental governance gap. This set the stage for individual company CSR efforts, unilateral industry codes of conduct, and the development of a broad spectrum of multi-stakeholder CSR standards.

Following UNCED, industry groups such as the World Business Council for Sustainable Development (WBCSD), the International Business Leaders Forum, CSR Europe, and Business for Social Responsibility formed to develop and promote CSR initiatives. It was not only corporations that were spurred towards initiating, developing, and implementing CSR standards. NGOs also played a key role.

During the 1980s, in the face of continuing evidence of corporate environmental abuses (e.g., the Bhopal chemical accident and the *Exxon Valdez* spill) and the mounting environmental effects of globalization (e.g., climate change, deforestation, depletion of the oceans, species extinction, and so on), environmental NGOs (ENGOs) recognized a need as well as an opportunity to develop new advocacy strategies. Instead of campaigning negatively against individual companies, they began working directly and collaboratively with corporations and industries to develop multi-stakeholder CSR standards. Their interest in working cooperatively grew not only out of concerns but also out of hopes that although increasingly powerful multinational corporations were a significant contributor to the worsening global environmental problems, these transnational firms, through their global supply chains, were also a potentially significant contributor to the solutions. Companies responded to the ENGOs to protect their reputations, avoid regulation, manage risks, and maintain their "social license to operate."[5]

The shift in ENGO strategy towards working collaboratively with rather than against industry was prompted not only by discouragement with the level of corporate commitments and what appeared to be ineffective business responsibility codes but also by a growing frustration with governments and with the inefficiency and ineffectiveness of state-based international processes. NGOs argued that neoliberal policies had brought forth "the competitive state," which was more focused on lowering trade barriers and creating financial incentives to attract mobile capital and achieve global economic competitiveness than on developing new international laws and multilateral agreements to halt environmental destruction (Barry and Eckersley 2005; Biermann and Dingwerth 2004; Eckersley 2004). Thus, in most cases, although the multi-stakeholder CSR initiatives leveraged international standards and agreements, they intentionally steered around government participation so as to avoid marginalizing the CSR outcomes. Industries supported this approach as they deemed governments to be inflexible and likely to stall the process.

Since UNCED, the result has been a rapid proliferation of CSR codes and standards developed by multinational firms and industry alone and/or in cooperation with civil society organizations, cutting across industry sectors and going beyond legal compliance and the reach of the state. Examples include company-specific codes of business conduct, unilateral industry codes of conduct,[6] multi-stakeholder industry-specific CSR standards, and cross-sector multi-stakeholder global CSR standards (Table 2.1).[7]

With increasing acceptance and adoption, these transnational codes and standards are becoming powerful governance mechanisms, and, as explained in the next section, many are gaining private rule-making authority.

The Institutionalization of CSR Initiatives

As noted earlier, the social roles and responsibilities of commercial entities have been debated for centuries. There is also a long history of governments permitting trades, industries, and professions to self-monitor their practices to ensure responsible conduct and fair play. So, is there really anything new about the present wave of CSR self-regulatory mechanisms? The answer is yes. Current CSR initiatives *do* constitute an important new governance phenomenon. In particular, an increasing number of CSR standards are becoming institutionalized – that is, they are gaining legitimacy and authority as private governance mechanisms that perform environmental policy functions similar to those of governments (Cashore 2002; Haufler 2001; Meidinger 1997, 1999).

Table 2.1

Corporate social responsibility codes and standards

CSR initiative	Description	Examples
Company codes of business conduct	Company statements of commitment to environmental and social responsibilities	Nike, Royal Dutch Shell, PepsiCo, Gap Inc., etc., global sourcing and worldwide codes of business conduct
Industry codes of conduct	Responsible business practices as defined by industry associations	Chemical Industry Responsible Care Program, Electronics Industry Code of Conduct, European Retail Code for Environmentally Sustainable Business, etc.
Industry-specific multi-stakeholder CSR standards	Responsible environmental and/or social business practices defined for a particular industry sector by a range of interested parties	Forest Stewardship Council (FSC) and Programme for the Endorsement of Forest Certification (PEFC) certification programs, the Marine Stewardship Council, the Equator Principles, RugMark, the Kimberly Process, fair trade certified coffee, bananas, etc.
Cross-sector multi-stakeholder CSR standards	Responsible environmental and/or social business practices that cut across all industry sectors, as developed by a range of interested parties	AA1000, the Global Reporting Initiative, the UN Global Compact, ISO 14000, etc.

CSR standards and codes represent a new governance approach as they have distinct design features compared with traditional examples of industry self-regulation. First, they are "non-delegated" – that is, they have not been formally initiated or sanctioned by the state but rather gain legitimacy through the acceptance of external actors. Second, most of these non-state initiatives are multi-stakeholder, developed by corporations and NGOs in partnership. And third, they are typically transboundary and multiscalar in nature, going beyond jurisdictional legislative constraints and operating in expanded political arenas that bridge local and global concerns.

Beyond this, certain CSR initiatives, such as certification programs, ecolabelling standards, and multi-stakeholder codes, are gaining private authority as they have specific features that constitute unprecedented self-regulatory governance capacity. For example, they have democratically designed, multi-stakeholder rule-making and adjudication bodies that operate under written constitutions. They also have independent audit processes to enforce compliance with a prescriptive standard. Because of this unique governance capability, as these CSR mechanisms gain acceptance by markets and society as well as governments, they are acquiring legitimacy and rule-making authority, essentially mimicking the policy role of public institutions.

The standards are gaining *market* acceptance among suppliers, manufacturers, distributors, customers, and consumers by leveraging the various industry supply chains. *Social* acceptance is occurring through open, ongoing multi-stakeholder participation. And the standards are achieving *governmental* acceptance through their co-regulatory design (e.g., incorporating legal compliance) and their potential to enhance state governance.

The distinct institutional strength of certification, eco-labelling, and multi-stakeholder codes is revealed through the three key aspects of governance – polity, politics, and policy. In terms of the polity, these private mechanisms are providing a new decision-making forum beyond the traditional state-centred political arena. With respect to politics, the private governance bodies are encouraging multi-stakeholder policy deliberation and increasing direct stakeholder rule-making responsibility. And with regard to policy, the private standards are establishing rules that not only reinforce legal requirements but also go beyond the law. Thus, as will be evaluated over the course of this book, the emergence of private environmental governance authority has significant implications for policy making, the traditional role of government, and overall state capacity to address local and global sustainability challenges.

Classifying Private Environmental Governance

From Government to Governance

Governance refers to a decision-making system that provides direction to an organization or society. Although there is no single standard definition of the term, in common political usage governance is ultimately about how to steer the economy and society towards the attainment of collective goals. Governance has therefore been synonymous with government, as democratic governments are vested with the constitutional political authority to make and implement rules (Stoker 1998). Today, however, new modes of governance have emerged that go beyond the traditional hierarchical model, in which state authorities exert sovereign control over society. Rather than government being at the centre of governing decisions, there are now new multicentric and private modes of networked and market-based governance, with government's role shifted towards greater steering, coordinating, and

Figure 2.1

Shifting modes of governance authority

Governance function	Co-regulatory governance system			
	Hierarchical	*Delegated*	*CSR co-regulatory*	*Non-delegated private*
Rule making	Command-and-control regulation	Market-based voluntary regulation Negotiated agreements	Multi-centric regulations CSR co-regulation	Private regulations CSR standards
Implementation (delivery)	Regulatory agencies	Industry self-regulation Policy networks Public/private partnerships	Public/private co-governance	Beyond compliance private initiative Multi-stakeholder self-regulation
Enforcement	Monitoring Compliance auditing	Shadow hierarchy Responsive regulation	Regulated self-regulation	Independent audits Public reporting Naming and shaming

Public authority ←——→ Private authority

facilitating through partnership arrangements and co-regulatory governance approaches. There is a shift from government to governance, whereby private actors participate to a greater degree in the formation and implementation of public policy and global governance mechanisms.[8] As illustrated below, not only are CSR multi-stakeholder standards a new governance mechanism but also government response to CSR constitutes a new *mode* of co-regulatory governance.

Environmental Governance Typology

The field of environmental governance is vast and includes varied terminology and definitions of traditional and new forms of governance and modes of governance authority. The typology presented in Figure 2.1 categorizes the field, first, along a continuum of public, private, and hybrid governance authority, and second, by regulatory function (rule development, implementation, and enforcement).[9] In particular, the typology highlights the unique case of non-delegated private governance and the emerging mode of multicentric co-regulatory governance. The combination of all four modes of governance constitutes a co-regulatory governance system. As explained later, it should be noted that these categories represent theoretical ideal types. Overlap and mixed modes occur in practice.

Hierarchical Governance

Hierarchical modes of governance concern the traditional bureaucratic, command-and-control style of direct government intervention through economic and social regulation. Legally binding standards are prescribed, policies and programs are implemented, and compliance is monitored through a government agency. With hierarchical governance, the state has central authority, makes the decisions, and enforces compliance.

Delegated Governance

Delegated governance refers to the state's handing off of governance functions to non-state actors. Governments maintain central authority but delegate certain self-regulatory responsibilities. This category of regulatory instrument constitutes the traditional forms of voluntary industry self-regulation. For example, in terms of rule making, rather than prescriptive, "hard law," command-and-control direct regulatory intervention, governments may use indirect approaches through "soft law," market-based voluntary instruments such as industry self-regulation and negotiated agreements and covenants, as well as informational tools and moral suasion.[10] The implementation of certain public services, provision of some public goods, and/or achievement of specific collective goals are formally delegated to the private sector through self-regulation, policy networks, and public/private partnerships. Governments focus less on "rowing" (direct delivery) and more on "steering" (enabling self-regulation) (Osborne and Gaebler 1993, 34; Rhodes 1996). Finally, regarding enforcement, government delegates responsibility for compliance to private actors under a *shadow of hierarchy,* meaning that they promote less coercive voluntary approaches but "move up" in terms of imposing direct intervention if there is industry non-cooperation.[11] In other words, the regulation is responsive (Ayres and Braithwaite 1992). Network governance scholars in particular refer to this as a form of *meta-governance* (government oversight of private networks).[12]

Non-Delegated Private Governance

As previously outlined, private governance refers to self-regulatory CSR codes and standards developed by private actors that have gained private rule-making authority. Unlike with delegated governance, CSR self-regulation occurs outside the realm of government sanction. Private governance concerns "non-delegated" private authority whereby the agenda, rules, implementation, and enforcement governance functions are carried out by private actors without necessary state participation and/or sanction. Implementation relies on voluntary corporate initiative to go beyond legal compliance to address social and environmental issues. Enforcement is achieved through independent third-party audits, transparency through public reporting, and "naming and shaming" by citizens, media, NGOs, and other firms.

Policy and governance analysts classify private governance under many conceptual labels. As a mechanism of industry self-regulation, it has numerous descriptors, such as "pure self-regulation," "unilateral self-regulation," or "multi-stakeholder regulation." Other terms include "corporate social responsibility" (CSR), "non-state market-driven (NSMD) governance," "non-state global governance," "private regulation," "private hard law," "civil regulation," and "corporate codes of conduct."[13] All of these labels highlight the private governance capacity of CSR codes and standards.

CSR Co-Regulatory Governance

The final mode of governance in the typology is co-regulation. This refers to a hybrid governance approach whereby regulations are specified, administered, and/or enforced through a combination of public and private rule-making systems.[14] Although similar to delegated public/private cooperative arrangements, CSR co-regulatory governance is multicentric in the sense that public and private policy authority coexist instead of authority residing solely in the state.[15] With CSR co-regulatory governance, therefore, private actors have rule-making authority rather than just policy influence.

Public authorities co-regulate CSR private rule making, implementation, and enforcement through enabling legislation, hard law regulation, and/or soft law approaches. The different means of meta-governing CSR include endorsing and enrolling in the private decision-making processes, enabling implementation, and/or mandating the uptake of private standards. Governments can also ignore, compete with, or block CSR efforts. The particular case of government *enforcement* of a private governance standard in a co-regulatory system is an example of "regulated self-regulation" (Knill and Lehmkuhl 2002; Schulz and Held 2004).

In summary, private governance mechanisms are a distinct policy instrument compared with traditional voluntary instruments because they have not been formally sanctioned by the state, and also because they have gained private rule-making authority. With private governance, private actors formulate the policy agenda, implement the rules, and oversee enforcement, while governments are placed in a lagging role of having to decide whether and how to respond. Government engagement in CSR therefore constitutes a new mode of governance. As is examined next, with CSR co-regulation public and private authority are coincident within a shared governance system.

Co-Regulatory Governance Systems

Political scientists have an ongoing debate as to whether the transformation from government to governance has constituted a "retreat of the state,"[16] a "hollowing of the government,"[17] or "governance without government."[18]

New governance scholars emphasize that instead of government retreating, traditional hierarchical forms of command-and-control intervention have been accompanied by other, more complex and fluid forms of governance that leverage the resources of private actors alongside state authority – examples of governance *with* government.[19] Rather than a hollowing of the state, there is a flux in regulation, with deregulatory and re-regulatory shifts occurring simultaneously (Ayres and Braithwaite 1992; Jordan, Wurzel, and Zito 2005; Utting 2005). Governance scholars refer to the rise of these many forms of regulation as the emergence of *regulatory capitalism*.[20]

Building on the new governance position, this section presents a conceptual tool for clarifying the mix of self-regulatory and co-regulatory tools of governance (beyond command-and-control regulation) that are emerging within co-regulatory governance systems. Before doing this, however, I review the traditional policy debate regarding the pros and cons of statutory intervention versus market-based voluntary self-regulation. This overview highlights the fact that although theorists generally paint a black-and-white distinction between these regulatory instruments, the public/private boundaries are in fact increasingly blurred. A comparative evaluation of several regulatory typologies reveals an increasingly complex array of unilateral, multi-stakeholder, delegated, and non-delegated self-regulatory and co-regulatory approaches that reflect varying public and private hybrid arrangements. This section concludes with an overview of the underlying objectives of an optimal co-regulatory policy mix.

Prescriptive versus Voluntary Policy Tools

Governments depend on markets for the efficient provision of goods and services that enhance social well-being. Markets depend on government rules to function efficiently and fairly. Achieving an optimum public/private balance of state intervention and market freedom is the subject of ongoing political debate. At one end of the spectrum are those who advocate "civic governance," whereby the state is required to intervene to protect the public good. Those on the opposite pole support an economic "consumer sovereignty" model of laissez-faire market dynamics and minimal government intervention. This fundamental political debate threads the environmental governance literature. Are sustainability goals best achieved by "hard law" legislated regulatory intervention or by "soft law" delegated voluntary approaches that leverage the power of the market to move firms towards better environmental performance?

Hard law and soft law approaches represent state-based regulatory and delegated voluntary mechanisms that range from high to low coercion (see Table 2.2). Assuming that policy instruments are substitutable, governments generally prefer the least intervention (lowest coercion) in order to maintain

Figure 2.2

Scale of policy coercion

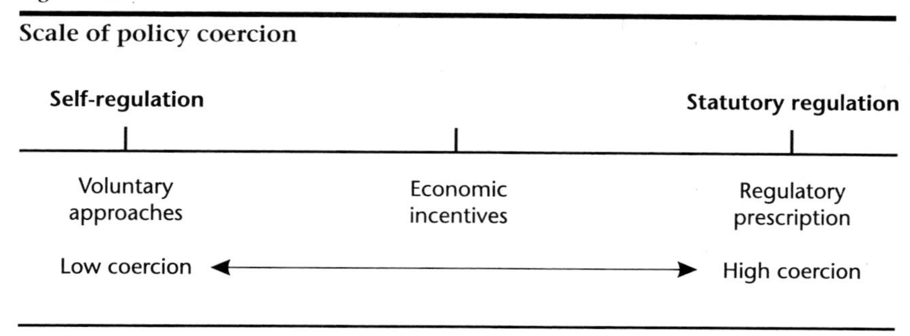

Table 2.2

"Hard law" versus "soft law" regulatory approaches

Type of intervention	Hard law (prescriptive)	Soft law (voluntary)
Setting standards	*Regulatory prescription* – traditional command-and-control regulation in which legally binding standards are prescribed	*Information* – influence constituents through the transfer of knowledge and the communication of reasoned argument and persuasion
Enforcing standards	*Economic regulatory instruments* – examples include pollution fees, emission taxes, and tradable permits to encourage firms to internalize environmental costs	*Voluntary approaches* – examples include industry self-regulation, codes, voluntary challenges, eco-labels, charters, co-regulation, covenants, and negotiated environmental agreements

legitimacy (policy acceptance). They then typically move up in level of coercion as necessary to overcome any social resistance to their policy goals and to achieve effective policy outcomes (Figure 2.2) (Phidd and Doern 1983).

Many studies have investigated the benefits and drawbacks of voluntary versus government-regulated approaches.[21] Traditional hard law regulation is generally criticized for being slow and expensive to develop, operate, and amend; for fostering adversarial relations; for dampening innovation and beyond-compliance behaviour; and for producing unintended outcomes. Voluntary approaches such as self-regulation are criticized for being difficult to apply, for being less rigorous in their performance requirements, and for their uncertain public accountability. Industry generally advocates the use of voluntary rather than regulatory approaches, as this avoids the imposition of inefficient regulation and offers policy direction while at the same time providing a flexible framework for innovation. Further, industry argues that self-regulation generates business process improvements and positive changes in corporate culture that are often hard to quantify. Policy scholars also argue that voluntary approaches can enhance efficiency and effectiveness by positioning the development and implementation of agreements in the hands of those closest to and most knowledgeable about the issues (Schulz and Held 2004).

Some analysts disagree with the generally held position that government regulations raise costs and encourage inefficiencies and competitive disadvantage. For example, Porter and van der Linde (1995) argue that properly designed environmental regulations can trigger innovations that can offset the costs of reducing the negative effect of operations on the environment, resulting in "enhanced resource productivity" (greater efficiencies) and making companies more competitive in the global market.

Research to assess the effectiveness of voluntary versus prescribed regulatory approaches has found that voluntary approaches as a stand-alone policy instrument generally fail to make substantial contributions to improved corporate environmental performance.[22] In particular, without the threat of government penalty, there is incentive for companies to have a "free ride," that is, to take advantage of benefits without participating and bearing costs. Research has also found that through processes such as negotiated voluntary agreements, governments can become "captured" by industry interests, thus compromising the achievement of performance targets.

Although most policy analyses emphasize the limitations of voluntary approaches on their own, these studies also highlight the fact that voluntary approaches, in fact, rarely occur as stand-alone policies. Rather, many voluntary approaches incorporate regulatory requirements and government oversight, and are seldom implemented in isolation from other policy instruments. This is demonstrated next with respect to the increasingly hybrid range of self-regulatory policy instruments.

Classifying Self-Regulatory Policy Instruments

From a political science perspective, self-regulation has traditionally referred to a state's delegation of regulatory powers to nongovernmental bodies. As explained above, self-regulation represents the low-coercion end of the scale of regulatory tools that governments can employ. Self-regulation is not a new concept. For example, there is a long history of governments sanctioning self-regulation in the broadcasting, communications, and financial sectors. Professional associations of engineers, actuaries, lawyers, medical doctors, accountants, and others continue to be largely self-regulated, although recently some have come under the increasingly watchful eye of the state. With neoliberal reforms, not only has state-delegated self-regulation become more prominent as a favoured environmental policy tool but also new forms of non-delegated CSR self-regulation have emerged. This has broadened the spectrum of self-regulatory approaches and created a varied landscape of regulatory terminology.

Here, I apply the previously introduced governance typology to the vast terrain of environmental regulation to sort and categorize the various theoretical definitions, distinctions, and approaches that scholars have employed to classify regulatory policy tools. Specifically, I review the self-regulation typologies developed by Haufler (2003), Knill and Lehmkuhl (2002), and Gunningham and Rees (1997).

The review begins with Haufler's classification of self-regulation based on rule-making authority, and then presents Knill and Lehmkuhl's typology of self-regulation in terms of policy implementation. It concludes with Gunningham and Rees's categorization of self-regulation based on the degree of government involvement in rule making and enforcement. The key point of the exercise is to illustrate how the three examples constitute different self-regulation classifications.

Rule-Making Authority

Haufler (2003) applies the criterion of rule-making authority to distinguish between four categories of regulation:

- Traditional regulation – Rules are developed, promulgated, and enforced by government.
- Industry self-regulation – The private sector develops standards and best practices on its own.
- Multi-stakeholder regulation – A variety of stakeholders, including non-profit groups, negotiate and develop a set of standards, a decision-making framework, and a process for achieving the standards.
- Co-regulation – Markets develop a standard and the public sector applies sanctions for non-compliance.

Responsibility for Implementation

Based on a consideration of the governance capacity[23] of public and private actors, Knill and Lehmkuhl (2002) offer a regulatory typology with respect to the locus of responsibility for the provision of public goods:

- Interventionist regulation (high public, low private capacity) – Overall responsibility for the provision of public goods lies with the state.
- Private self-regulation (high private, low public capacity) – Provision of public goods by private actors.

- Regulated self-regulation (high public, high private capacity) – Cooperative public/private governance.
    - Private actors participate in policy making and implementation.
    - Competencies are delegated to private organizations.
    - Regulatory frameworks for private self-regulation are cooperatively developed.

Government Involvement in Rule Making and Enforcement

Finally, Gunningham and Rees (1997) combine aspects of the previous typologies by distinguishing self-regulatory approaches based on the degree of government involvement in both rule-making authority and enforcement:

- Voluntary self-regulation – Rule making and enforcement by the firm or industry itself, independent of direct government involvement.
- Mandated full self-regulation – Rule making and enforcement privatized but sanctioned by government, which monitors the program and, if necessary, takes steps to ensure its effectiveness.
- Mandated partial self-regulation – Privatization of either rule making or enforcement but not both.
    - Public enforcement of privately written rules, or
    - Government-mandated internal enforcement of publicly written rules.

In summary, this review of regulatory typologies highlights two main findings. First, the definitions of self-regulation vary according to the extent of government engagement; the stage of the policy cycle (rule making, implementation, or enforcement); the degree of corporate and/or NGO involvement and authority; and the focus on individual firms versus industries. Second, there is a lack of definitional consistency in the policy literature; that is, there is great variation in the regulation terminology.

To a large degree, the definitional confusion stems from the first finding. Rather than dichotomous "pure forms" of either self-regulation or government regulation, there is now a continuum of hybrid arrangements that reflect varying degrees of government involvement and different public/private arrangements of rule-making authority and of implementation and enforcement responsibility. This blurring of public and private boundaries is characteristic of emerging co-regulatory systems of governance. The conceptual map presented next helps to clarify the various categories of governance instruments within a co-regulatory policy mix.

Co-Regulatory Policy Mix

> Not only do public policy choices and public policy networks influence the emergence of non-state authority,

> but it is now increasingly clear that private authority is influencing the emergence of new public policy initiatives, including their content and instrument design. (Cashore et al. 2009, 230)

As Cashore and colleagues note, there is an increasingly dynamic and synergistic interaction between public and private environmental rule-making systems. Peter Utting, the deputy director of the United Nations Research Institute for Social Development, describes this mixing of CSR and traditional policy tools as a re-regulatory trend towards "articulated regulation" – a coming together of different regulatory approaches in ways that are complementary and synergistic (Utting 2005). Gunningham and Grabosky (1998, 15) further explain that "recruiting a range of regulatory actors to implement complementary combinations of policy instruments, tailored to specific environmental goals and circumstances, will produce more effective and efficient policy outcomes."

Fundamentally, as multi-stakeholder CSR initiatives gain rule-making authority, there is increased interaction with public policy, and a growing prospect for the co-regulation of these private environmental governance mechanisms to enhance government policies and programs. The mixing and temporal sequencing of various public, private, and co-regulatory instruments at the different stages of the policy cycle constitute a co-regulatory governance system. This is illustrated in Figure 2.3 by the overlapping co-regulatory governance circle.

The circle encompasses traditional command-and-control regulation, delegated self-regulation, non-delegated self-regulation, and CSR co-regulation. Specifically, Figure 2.3 shows how the various tools of governance are situated according to the degree of public versus private rule-making authority, and the extent to which they are state- versus market-driven. For example, command-and-control regulation represents a traditional state-centric, government-driven policy instrument. Delegated self-regulation is also a traditional policy tool, with authority continuing to reside with the state but with self-regulatory policy responsibilities delegated to the market. Non-delegated self-regulation and regulated self-regulation constitute co-regulatory governance instruments that leverage private authority. And, as already noted, the overall mix of these various regulatory instruments constitutes a co-regulatory governance system.

Although the boundaries between the cells in Figure 2.3 are shown as theoretically distinct, in practice they are overlapping and porous. For example, regulated self-regulation combines command-and-control regulation and non-delegated self-regulation – that is, a prescriptive legislated requirement to comply with a voluntary CSR standard. As well, the categorization of non-delegated self-regulation as a purely non-state market-driven (NSMD)

Figure 2.3

Co-regulatory policy mix

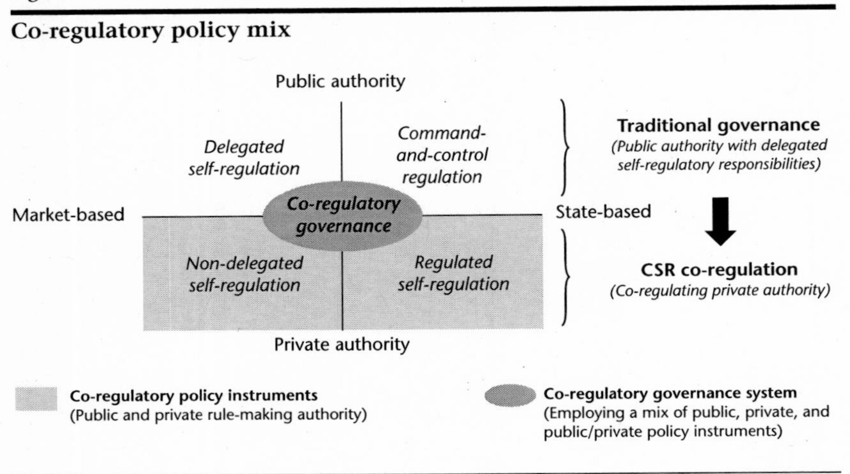

mechanism is only a partial account. As will be demonstrated later in this book, in the case of forest certification these private governance CSR mechanisms comprise public and private rules, and are developed and delivered with varying degrees of government engagement. Ultimately, non-delegated private self-regulatory regimes overlap with public governance and rely on enabling legal frameworks and overarching legislative oversight, and civil society actors and the state (not just markets) have also been drivers. Thus, NSMD mechanisms are not purely private but rather co-regulatory instruments that operate within co-regulatory governance systems.

Beyond the challenge of sorting the definitional categories of new modes of governance and co-regulatory policy instruments, there is also the question of how to combine the various governance tools so as to achieve an optimal mix that maximizes the strengths while minimizing the weaknesses of the various policy approaches. Although there is a growing body of literature on the "new tools of governance" that concerns the development and application of optimal mixes of direct and indirect policy instruments in response to new multicentric, collaborative modes of governance, there is still no established theory of regulatory choice in terms of optimizing the co-regulatory mix of state, market, and NGO-led regulatory mechanisms.[24] In the absence of a theory, three fundamental concepts can guide instrument selection: "responsive regulation," "minimal sufficiency," and "smart regulation." These are outlined below.

Ayres and Braithwaite (1992) argue for responsive regulation and minimal sufficiency (namely, that regulation will be more effective the more sanctions can be kept in the background and regulation transacted first through moral suasion – that is, the regulation is responsive to corporate behaviour), allowing for corporate virtue before turning to coercive enforcement. They explain that to achieve the best behavioural outcomes, the social and environmental responsibilities of the firm (CSR) should be appealed to before any regulatory intervention, and they note that the trick of successful regulation is to establish a synergy between punishment and persuasion so as to encourage CSR self-regulatory commitment (Ayres and Braithwaite 1992, 53).

In order to illustrate their point, they developed a two-dimensional "enforcement strategy pyramid" depicting how governments are most likely to achieve their goals with business through self-regulation and by communicating to industry their willingness to escalate their regulatory strategy to another level of intervention in the pyramid if industry fails to cooperate. The pyramid progresses from self-regulation to enforced self-regulation, to command regulation with discretionary punishment, to command regulation with non-discretionary punishment.

Following on from Ayres and Braithwaite, Gunningham and Grabosky (1998) introduced the concept of "smart regulation," or expanding the traditional use of command-and-control regulation to include an alternative, innovative, and more dynamic mix of policy instruments that harness not just governments but also businesses and third-party NGOs. Specifically, they argue that in addressing complex environmental issues, the use of multiple instruments and a broader range of regulatory actors will produce better environmental regulations. In other words, "smart regulation" is imaginative, flexible, and pluralistic in approach.

To illustrate the concept, they expanded the Ayres and Braithwaite pyramid to include three dimensions. The three sides reflect the possibility of rule making and enforcement (coercion) not just by government but also by business and third-party NGOs (such as environmental advocacy groups) moving up each of the three faces of the pyramid. The figure depicts the coordinated use of a number of different instruments across a number of different governance spheres – in other words, a co-regulatory governance system.

In terms of achieving "optimal" smart regulation, the authors identify combinations of inherently complementary and non-complementary policy instrument mixes. They also highlight the importance of sequencing in introducing instrument combinations.[25]

Specifically, they stress five key principles of design in a "smart" regulatory mix:

- prefer policy mixes incorporating a broader range of instruments and institutions

- prefer less interventionist measures by applying instead the principle of low interventionism
- ascend a dynamic instrument pyramid to the extent necessary to achieve policy goals
- empower participants who are in the best position to act as surrogate regulators by applying the principle of empowerment
- maximize opportunities for win/win outcomes.

In summary, the regulatory enforcement pyramids highlight the necessary dynamic between government, business, and nongovernmental third parties within co-regulatory governance systems. The concepts of responsive regulation, minimal sufficiency, and smart regulation also stress the goal of achieving a mix of policy instruments that balance penalty and reward – a combination of coercive and voluntary approaches that enable corporate social responsibility initiative.

Evaluating Private Environmental Governance

As discussed in the previous sections, co-regulation is a means of combining the strengths of command-and-control regulation (credibility, accountability, enforceability, and greater performance rigour) with the strengths of self-regulation (speed, flexibility, and innovation) while avoiding the drawbacks of each. An optimal policy mix achieves an effective balance of prescriptive and voluntary regulations. Because private governance CSR standards are not formally delegated by the state, they present an additional co-regulatory challenge compared with other self-regulatory instruments. Specifically, in the absence of formal state sanction, to what extent are the private rule-making bodies and standards legitimate? In other words, a critical question for governments is not just *how* private governance should be incorporated in the policy mix but *whether* CSR should be co-regulated. Ultimately, should governments endorse or compete with private rule-making authority? A significant contributing factor to this decision relates to whether in co-regulating CSR governments are upholding the principles of good governance.

The term "good governance" is most commonly employed in the global realm to emphasize the importance of a capable state operating under the rule of law, and the achievement of democratic core values such as participation, representation, and political contestation (Kaufmann, Kraay, and Mastruzzi 2007). The traditional evaluation of good governance with respect to state-based policy tools and regulatory approaches includes the consideration and weighting of several dominant instrument selection criteria, such as effectiveness, efficiency, legality, and democratic values.[26] It is not clear, however, whether it is appropriate to apply these same criteria concerning the ideal of democratic governance on the part of government to private CSR non-state governance mechanisms.

Although a definition of good governance with which to assess non-state private governance standards in either domestic or global realms has not yet been established, several applied efforts have emerged in an attempt to define and assess CSR initiatives. For example, the World Bank has established criteria to assess CSR codes and standards in terms of their potential to contribute to public governance. The International Organization for Standardization (ISO) has launched ISO 26000, an international standard that provides a global CSR definition and benchmark. The International Social and Environmental Accreditation and Labelling (ISEAL) Alliance has developed a Code of Good Practice for assessing the credibility of voluntary environmental and social standards. An increasing number of individual governments are also developing criteria by which to judge the credibility of CSR standards for inclusion in public procurement policies.

In the absence of a single, established framework, CSR private governance analysts have generally adopted three assessment criteria that overlap with democratic principles and state-based definitions of good governance. These include consideration of the legitimacy, accountability, and effectiveness of the private rule-making systems.

Legitimacy refers to the acceptance and justification of shared rule by the affected community – having the consent of the governed, so to speak.[27] Beyond assessing the acceptance and uptake of the private standards, legitimacy also relates to the accountability and effectiveness of the governance mechanism. Accountability concerns the justification of a decision maker's actions vis-à-vis the affected parties (stakeholders): "those who are assigned responsibility are obliged to answer for their performance."[28] Accountability is commonly evaluated through "input-oriented legitimacy" criteria, including inclusiveness (representative participation), responsiveness (openness to stakeholder input), and transparency (reliable information and communication). Finally, effectiveness is a form of "output-oriented legitimacy" and fundamentally concerns the problem-solving capability of the governance mechanism.[29] In addition, measures of policy effectiveness include output, outcome, and impact criteria.[30]

Employing these criteria, we see that perspectives on the good governance potential of private governance systems fall broadly into two camps: traditional "statists" and "new governance" theorists. Table 2.3 provides a summary of these perspectives.

Statist

The statist perspective assumes that political decision making resides with the state and that authority is autonomous and zero-sum rather than

Table 2.3

Theoretical perspectives on private environmental governance

Governance characteristic	Definition	Traditional statists	New governance
Legitimate authority	Having the consent of the governed	*No* – Unelected and therefore lacking democratic consent and accountability. Private standards have influence but not legitimacy, as political authority resides only with the state.	*Yes* – Increasing adoption demonstrates acceptance and legitimacy. Authority is derived from different forms of legitimacy (e.g., cognitive, moral, pragmatic legitimacy). Sovereignty is relational rather than insular.
Accountability	Inclusive, responsive, and transparent	*No* – Uncertain due process, unbalanced representation, fragmentation of accountability channels, and a lack of checks and balances.	*Yes* – Societal trust is equivalent to constitutional checks and balances or formal contract. Inclusive multi-stakeholder process (street-level democracy) and transparency through third-party audits.
Effectiveness	Relevant policy with high uptake (output) + behavioural change (outcome) = problem resolution (impact effectiveness)	*No* – Uneven uptake and "cherry picking" among a confusing array of standards. Demand-side issue of increasing consumption not addressed. "Fox" is left "guarding the henhouse."	*Yes* – Uptake is increasing and the corporate actors causing the problems are contributing to solutions. Innovative and responsive, with transnational reach through global supply chains.

potentially expanded through complementary public and private governance capacities. Statists argue that private environmental governance has neither input nor output legitimacy, as consent is not granted through formal state sanction or through an electoral process, and outcome objectives are partial to a stakeholder subset rather than the overall citizenry. Private standards are deemed to merely have influence as opposed to authority. Further, statists argue that private governance mechanisms lack accountability due to fragmented rather than centralized authority, and that they are not an effective means of governance because their voluntary nature means that there is partial, patchy, and uncertain uptake, with essentially "the fox left to guard the henhouse."[31]

New Governance

New governance theorists, on the other hand, assume a plurality of authority (sharing of various aspects) among state and non-state actors. State sovereignty is no longer about autonomy and insular governance but rather post-sovereign interdependence and relational governance. Proponents of new governance argue that private environmental governance mechanisms are legitimate as consent for these mechanisms is achieved by the evaluation of external actors and demonstrated by the increasing adoption of the standards – that legitimacy is gained by means beyond state sanction, such as forms of moral, cognitive, and pragmatic legitimacy (Bernstein and Cashore 2007; Cashore 2002). Accountability is gained through multi-stakeholder processes and third-party audits; in fact, private standards improve public accountability by providing a form of "street- level" democracy. Finally, the new governance perspective views private environmental governance as encouraging effective governance because private standards leverage the human, technical, and financial resources of transnational corporations so that the key actors involved in the creation of global issues are also directly involved in finding solutions to the issues (Vogel 2006).

Thus, as a new mode of governance, private environmental governance mechanisms such as multi-stakeholder CSR standards can be viewed as a positive development because they directly engage multinational corporations and their global supply chain partners, who are key contributors to sustainability issues and solutions. By increasing corporate and civil society responsibilities, private governance also provides additional resources (often with transnational reach) that consequently increase domestic and global problem-solving capacities. Nevertheless, since these private standard-setting mechanisms are established without formal state sanction (and are therefore not democratically accountable in the traditional sense), the encouragement of private authority may constitute an erosion of state sovereignty and, further, a subversion of the democratic process. Ultimately, the statist and new governance perspectives highlight the strengths and weaknesses of private environmental governance and the importance of baseline public capacity to effective CSR co-regulation.

Despite acknowledgment of the importance of public governance to CSR, there has been little theoretical or empirical research regarding how private governance standards interact with public policy, and how and why governments are responding. CSR governance research has focused for the most part on the limits and potential of CSR in developing regions, where private standards could possibly fill a governance gap.[32] The extent to which private environmental governance standards constitute an effective policy tool and a good governance mechanism in industrialized regions with strong public institutions remains largely unexplored. To facilitate empirical investigation in this regard, the following regulatory scale (adapted from a World Bank CSR tool) can be used for mapping and evaluating the range of government responses to such private standards.

Public Sector Role in Co-Regulating CSR

Although the domestic policy role of private environmental governance is uncertain, there is an emerging broad consensus among public and private actors that governments *can* do much to enhance the effectiveness of CSR standards through supportive, coordinated, and enabling policies, and by showing strong political leadership on CSR.[33] Governments and NGOs are also recognizing that CSR standards can play a role in supplementing public policy. Consequently, in many jurisdictions governments have responded eagerly to CSR.[34] Here, I assess the key rationale for government engagement in CSR, and outline the spectrum of indirect to direct approaches to co-regulating CSR.

Rationale for CSR Co-Regulation

Just as companies need to understand the business case for CSR, governments need to understand the public policy case for encouraging private governance authority. Private CSR standards represent both an opportunity and a potential threat to a government's policy agenda. On the one hand, private governance presents an opportunity to leverage corporate resources, lessening the government's regulatory costs and potentially achieving overall efficiency benefits.[35] On the other hand, there is uncertainty regarding the content and uptake of voluntary CSR standards, and there is a danger that by placing sustainability decisions in the hands of private actors rather than democratically elected governments, public authorities may lose control over the local policy agenda.

Thus, the challenge for governments is to find means to ensure that private regulations complement rather than supplant public policy goals. Specifically, as with any voluntary initiative, the state may need to intervene in order to ensure that private governance standards achieve three fundamental objectives: to operate in the public interest, to be effective in achieving their purported social, environmental, and economic goals, and to have credibility in the eyes of the public. In addition, the rationale for government engagement with CSR may necessarily include considerations of national competitiveness, the potential to leverage private resources, popularity with the electorate, and win/win business and social opportunities in enabling CSR.

Government Role in CSR

Although CSR studies to date have largely focused on the corporate motivation and business case for beyond-compliance initiative, research findings are beginning to emerge regarding public governance and CSR. In particular, as a consequence of concerns that the global CSR agenda is being dominated by northern economies, is focused on large enterprises, and is limited to voluntary business activity only, several industry, academic, and policy research organizations have recently turned their attention towards understanding the role of government in creating an enabling environment to "scale up" corporate responsibility efforts in developing and transitioning economies.[36] This research has, in turn, identified a large knowledge gap with respect to how and why governments are engaging in CSR and leveraging voluntary CSR standards to carry out their policy objectives.

Studies that have evaluated the role of government in CSR have identified a spectrum ranging from direct to indirect responses depending on the specific context. For example, from its case studies on government's role in CSR in developing countries, the World Bank's Foreign Investment Advisory Service (FIAS) Corporate Social Responsibility group identified a range of government approaches towards enabling CSR, including mandating, facilitating, partnering, and/or endorsing CSR initiatives.[37] Similarly, the ISEAL Alliance's recent comparative case study investigation of governmental use of voluntary standards identified three institutional arrangements:

- users – governments that have a direct relationship with the voluntary standards systems
- supporters – governments that provide incentives related to affiliation with a voluntary standards system
- facilitators – governments that provide a favourable policy environment or resources to facilitate the development of a multi-stakeholder voluntary standard.

Other dimensions of government CSR response include a neutral role of just observing, or an unsupportive role of prohibiting or competing with the private standards.

Expanding on the these frameworks, the various forms of government response to CSR can be situated along a "spectrum of engagement" from observing to indirect cooperation, to more direct enabling, endorsing, mandating, or blocking of self-regulatory standards and other CSR tools. Government CSR positioning (whether direct or indirect intervention) may also vary in terms of governance function, whether at the rule-making, implementation, or enforcement stages. Incorporating these two dimensions, the spectrum of government's co-regulatory role in CSR is mapped in Figure 2.4.

Figure 2.4

The spectrum of government's role in CSR

Spectrum of government CSR intervention

Indirect ⟷ Direct

	Observe	Cooperate	Enable	Endorse	Mandate (or block)
Rule making					
Implementation					
Enforcement					

The categories of government response to CSR include:

- observing – watching rather than interfering; leaving CSR development, implementation, and enforcement to the market
- cooperating – providing informational and/or technical assistance
- enabling – facilitating CSR initiatives by providing incentives
- endorsing – setting an example of best practice by adopting CSR standards in administrative policies (e.g., public procurement standards)
- mandating or blocking – establishing the CSR initiative as a legislated requirement (e.g., regulated self-regulation) or directly intervening to block its development and implementation.

The indirect and direct ends of the spectrum represent essentially two schools of thought. On the indirect end, the role of government is perceived to be that of supporting the CSR culture among enterprises and facilitating the development and implementation of the private standards through informational and incentive-based "soft" approaches. On the extreme opposite, direct end of the spectrum, the role of government is seen as that of establishing "hard law" regulations with respect to CSR standards. Depending on how government positions itself, from passive observer to mandating or blocking CSR standards, there are a number of mixed public/private co-regulatory approaches the state can employ to encourage effective CSR. Table 2.4 provides examples of the range of CSR co-regulatory roles at the various policy stages.

Proponents of direct approaches believe that co-regulating CSR will speed up effective implementation and prevent CSR from becoming simply

Table 2.4

Government role in CSR co-regulation

Aspect of co-regulation	Indirect role	Direct role
Rule making	Provide resources and technical guidance to standards development	Participate in negotiation of rules and/or establish enabling baseline regulatory framework
Implementation	Provide information and incentives, and remove any administrative or policy barriers	Adopt CSR standards in public administration and public procurement policies
Enforcement	Threaten to mandate or obstruct CSR standards	Mandate or block CSR standards

corporate propaganda and a "greenwash" marketing tool.[38] Furthermore, state intervention will establish boundaries around expectations regarding CSR and ensure greater fairness and more even uptake of CSR rules.[39] The European Commission has clearly positioned itself as a proponent of voluntary rather than prescriptive CSR intervention. European Union (EU) member countries have therefore adopted soft approaches to achieving CSR objectives by promoting stakeholder dialogue and public/private partnerships, enhancing transparency and credibility of CSR practices and instruments, raising awareness, increasing knowledge, disseminating and awarding best practices, and ensuring a link between sustainable development objectives and public policies (European Commission 2007, 3).

Canadian government consultation with industry leaders on the question of government's role in CSR also reveals an expectation by industry for indirect facilitative rather than direct interventionist co-regulatory government response to CSR (Natural Resources Canada 2004b). Specifically, company leaders recommended that the Canadian government support CSR initiatives by acting as a role model, disseminating best practices, recognizing companies that are leaders in CSR, providing incentives, and developing programs to support company CSR efforts.

Given what appears to be a consistent, stated government preference in developed countries for minimal CSR intervention, it is curious that in the case of forest certification, governments have gone beyond indirect approaches to demonstrate direct CSR co-regulation, including mandating this private governance mechanism. This puzzle is explored in the next chapter.

Summary

Corporate and multi-stakeholder CSR initiatives that are voluntary and that go beyond the law are emerging across global industry sectors, from fisheries, forestry, and mining to tourism, electronics, and retail. Many of these private rule-making efforts are becoming institutionalized as private environmental governance mechanisms that behave like public policy, and thus are playing an increasingly important role in shaping corporate sustainability behaviour.

Transnational CSR standards constitute an unprecedented form of private governance authority and can therefore be classified as a new mode of non-delegated governance that is distinct from the traditional state-centric hierarchical and delegated regulatory approaches. Although in theory these policy instruments can be grouped into separate categories, in practice they overlap. This occurs within an overall co-regulatory governance system.

The challenge for the state in addressing CSR is to determine the optimal public/private co-regulatory policy mix. The task is made more difficult by the uncertainties of defining and assessing CSR effectiveness. CSR standards are voluntary, dynamic, and often related to improving a range of business management processes rather than a single bottom-line performance result. In the absence of a standardized effectiveness measure (such as profit), good governance criteria regarding the legitimacy, accountability, and outcomes of private environmental governance mechanisms can provide helpful proxy indicators. Governments can use these measures to establish their optimal response to CSR along a spectrum from indirect facilitation to direct engagement at the development, implementation, and enforcement stages of the policy cycle.

Overall, private environmental governance CSR mechanisms represent an increasing and important regulatory challenge for the state. On the one hand, CSR has the negative potential to subvert government policy-making authority. On the other hand, through strategic government response, these standards can positively complement and supplement public regulatory functions. In order to move beyond the general theory towards an understanding of the applied empirical reality of the co-regulatory dynamic, we turn now to the specific case of forest certification.

3
Government's Role in Forest Certification

Forest certification is a CSR private environmental governance mechanism that has emerged in recent years as an international standard of proof that forests are sustainably managed. Independent of formal state sanction, multi-stakeholder certification organizations are carrying out traditional state functions by developing, implementing, and enforcing private rules for forest management. Certification works by leveraging global supply chains, linking customer demand for certified forest products with producer supply. If a forest manager/owner is found by an independent, nongovernmental certification body to be managing its forest in conformity with a set of internationally and regionally accepted sustainable forest management (SFM) principles and criteria, a certificate is issued that enables the forest owner to bring its forest products to market as certified. The objective of forest certification is to encourage forest producers around the globe to voluntarily adopt progressive SFM practices.

Compared with other CSR standards, participation in certification has been relatively enthusiastic and market demand continues to grow. Governments, industry, and small forestland landowners have signed on, but uptake has been uneven. Although certification was initiated to address a regulatory gap in terms of massive deforestation in southern tropical regions, it has instead been adopted as a supplementary forest policy tool in developed countries. Approximately 90 percent of the world's certified forests are located in industrialized nations in the Northern Hemisphere. Whereas certification struggles to achieve scale in developing regions, its rapid adoption in highly regulated global forest-product-producing countries provides a window of research opportunity to evaluate the interaction of public and private rule-making authority, and emerging co-regulatory governance arrangements.

Assuming that forest certification rules align with the domestic forest policy agenda in developed countries, certification should in theory not only lessen a government's forest management regulatory costs but also lead to enhanced SFM policy and outcomes. The challenge for governments is to determine the optimal response to certification so as to enable private rule-making innovation and enhance the potential forest governance benefits while also maintaining state forest policy sovereignty. Governments *are* engaging in certification through a range of approaches, but the optimal government co-regulatory response is unclear.

This chapter follows a progression similar to that in Chapter 2 in evaluating the nature and importance of the public/private co-regulatory dynamic between traditional state authority and private regulation in the case of forest certification systems. Here, I argue that forest certification is fundamentally a co-regulatory governance mechanism, relying on state authority, a baseline regulatory framework, and government support. The chapter defines and assesses the various forms of coexistence between forest certification and traditional regulatory and self-regulatory policy instruments within hybrid co-regulatory forest governance systems. Finally, in conclusion I argue that, depending on conditions within the domestic forest regime, there are a range of indirect to direct approaches by which governments can co-regulate certification in parallel with traditional regulation, with the potential to enhance overall forest governance.

The Emergence and Evolution of Forest Certification

The War in the Woods and Tropical Deforestation

Throughout history, forests have been critical to human settlement, providing the essential food, fuel, and shelter for expanding human populations. With rapid industrialization over the past two centuries, however, accelerating demand has resulted in a dramatic decline in forests around the world, with fifty-four countries losing as much as 90 percent or more of their forest cover (Millennium Ecosystem Assessment 2005, 587). There are some positive trends; for example, over the past fifteen years, global forest cover has been relatively stable and has even been increasing in some areas, such as Western Europe, North America, and China.[1] Deforestation and forest degradation continue to increase in several "hot spots," however, specifically, the tropical regions of South America, Asia, and Africa.[2] These ecologically rich tropical forests are disappearing at a rate of 12 million hectares per year, largely through conversion to more economically lucrative agricultural crops (such as soy and oil palm), cattle ranching, and fast-growing timber plantations. At present, primary forests comprise one-third of the global forest area. These last remaining frontier forests include not just the Brazilian Amazon and Southeast Asian tropical rainforests but also the temperate rainforests of the Pacific Northwest coast of North America and the vast northern Russian and Canadian boreal forests. Halting tropical degradation and deforestation and protecting ancient intact primary forests are among the top priority global environmental advocacy concerns.[3]

In northern temperate regions during the 1970s and 1980s, public concerns greatly increased over the protection of non-timber forest values such as wildlife, recreation, aesthetics, and cultural values. At the same time, scientific research confirmed the essential soil, water, biodiversity, and climatic ecological services provided by forests. In response, many governments revised their forest laws to consider conservation values, including forest preservation and forest restoration objectives, and temperate forest cover expanded overall. However, with unchecked global demand for paper and forest products and increasingly mechanized forest harvest practices, timber production continued to expand and forest conflicts also increased. For example, forest management in North America in the 1980s and 1990s was largely characterized by a "war in the woods" – bitter political battles between governments, the forest industry, and environmental organizations over clearcutting harvest practices and the protection of old-growth forest.[4]

Such conflicts revealed a fundamental challenge to forest governance: that of achieving a balance among a range of interests and shifting forest values. Governments and forest companies gradually realized that sustainable forestry was not just about maximizing timber production but also about achieving ecological preservation and respecting social and cultural forest values. In the early 1990s, in response to societal pressures, major producer countries such as Canada, Sweden, and Finland further revised their forest laws to balance timber production and ecological objectives. Governments also initiated voluntary and collaborative approaches to increase corporate responsibility and encourage greater stakeholder dialogue and engagement in forest governance decisions.

At the same time, society awoke to the global crisis involving the rapid loss of primary tropical forests and the linkage between deforestation and climate change. Satellite images showing vast clearcut forest areas from space set off global alarm bells. During the 1980s, environmental nongovernmental organizations (ENGOs) achieved greater global coordination and began to demand international laws to address weak tropical forest governance. Specifically, ENGOs wanted to increase the accountability of multinational corporations such as Aracruz Celulose, Mitsubishi, and McDonald's Corporation that were deemed to be taking advantage of minimal forest laws to ravage tropical forests (Dauvergne 1997, 2001, 2005). The results of state efforts were disappointing, however, and ENGOs, pessimistic about the prospects for an international forestry agreement, turned instead to developing a private governance mechanism – forest certification.

The Failure to Establish a Binding Global Forest Convention

Despite repeated attempts over the past fifteen years, the global community has been unable to negotiate a *legally binding* agreement on forests.[5] This is not only because establishing international law is inherently difficult but also because forests present a unique global environmental governance challenge.

It is important to protect forests because they provide essential local ecological services and resources as well as provide vital global benefits. They are commonly described as "the lungs of the planet" and giant "greenhouse gas sinks," taking in and storing vast amounts of carbon dioxide and emitting oxygen as the by-product of photosynthesis. Forests also constitute some of the last remaining tracts of wilderness on the planet, providing critical habitat for the growing list of threatened and endangered species.

Forests are fundamentally difficult to govern because they represent both a public good and a private resource. For example, carbon sequestration and forest habitat biodiversity are global public goods – no one pays and everyone benefits. Timber, however, is also a private resource under sovereign state authority. Governments have historically managed their forests to maximize resource benefits in the national interest, for example, to provide material for navy ships, state building construction, and fuel energy, and to generate jobs and capital. Reconciling national interest with global public benefit is thus a key global forest governance challenge. As forest governance scholars Lipschutz and Rowe (2005, 110) explain, "because forests are in effect private resources whose market (*timber*) value is easily determined, there is considerable reluctance to give away any of that value in the pursuit of some poorly-defined global good whose benefits are widely spread and difficult to quantify."

Strong national interest has been a fundamental roadblock to the establishment of a global forest convention. Southern developing states and northern industrialized countries have simply not been able to agree on the balance between economic, social, and environmental considerations to facilitate global forest products trade, protect domestic interests, and also prevent deforestation and forest degradation.

Intergovernmental attempts to protect the global public forest good through a global convention began in the 1990s. There were two unsuccessful rounds of negotiations; the first took place at the preparatory meetings prior to the 1992 United Nations Conference on Environment and Development (UNCED), and the second took place between 1995 and 1997 under the auspices of the Intergovernmental Panel on Forests. At the 1992 UNCED in Rio de Janeiro, northern countries argued for a global responsibility approach through a legally binding convention, whereas southern states wanted sovereign discretion. G77 (Group of Seventy-Seven) countries such as Malaysia, India, and Brazil strongly opposed any kind of global forest regulation, essentially viewing it as an attempt by industrialized countries to gain control of tropical forests without any form of compensation. Thus, instead of a legally binding agreement, the UNCED negotiations resulted in the adoption of the "Non-legally Binding Authoritative Statement of Principles for a Global Consensus on the Management, Conservation and Sustainable Development of All Types of Forests" (UN Forest Principles) as well as a general chapter of Agenda 21 on "Combating Deforestation." Although governments achieved consensus on a broad set of SFM principles, the non-binding UNCED outcome essentially reinforced the status quo.

NGOs, frustrated and disappointed with the failed state-based efforts leading up to Rio and at the 1992 conference, changed their position from supporting to opposing a legally binding convention.[6] A September 2004 joint NGO statement to the United Nations Forum on Forests (UNFF) summarized their lack of support for a binding global forest convention. It stated that the convention would reinforce the status quo and fail to address the underlying causes of forest loss, which included lack of recognition of indigenous peoples' rights, unsustainable consumption and production patterns, and unsustainable financial and timber trade flows.

Thus, sparked by the failure of governments to cooperatively agree on a binding forest convention, the World Wildlife Fund (WWF), along with other global ENGOs, disengaged from the formal international negotiations and refocused their efforts on a separate, already partially formulated private, multi-stakeholder global forest governance certification process targeted directly at corporations. The Forest Stewardship Council (FSC) forest certification scheme was then established in 1993.

Competing Forest Certification Programs

The Forest Stewardship Council

Building on earlier certification efforts by groups such as the Woodworkers Alliance for Rainforest Protection and draft standards prepared by the Rainforest Alliance and WWF, environmental and civil society NGOs along with invited industry officials formed the FSC to address a governance gap in global forest management, specifically, the inadequacy of governmental response to the deforestation and degradation of tropical forests and the loss of the remaining pockets of ancient northern temperate and boreal forest. Governments were explicitly excluded. The intent was to work directly with progressive companies to encourage the adoption of beyond-compliance SFM practices. Having recently observed the difficulties and failure at UNCED, the FSC founders wanted to make sure that governments did not unduly influence, stall, or marginalize the FSC process. (The UNCED forest principles were, however, included in the development of the FSC standard.) With the exception of governments, membership in the FSC was open to all individuals and organizations with a stake or interest in forest governance. They could be chosen to participate in one of two equally balanced economic or social/environmental chambers that comprised the General Assembly. (In 1996, this was revised to three chambers: economic, environmental, and social.) The FSC also ensured that northern and southern groups were represented equally within each chamber. And finally, the FSC encouraged similar multi-stakeholder engagement within the national and/or subnational FSC regional processes that reported to the FSC International General Assembly. All FSC bodies operated on the basis of consensus decision making.[7]

The FSC participatory structure was designed to address shortcomings that had been observed in the international forest negotiation process, including the exclusion of civil society actors, the overweighting of industrialized regions relative to developing regions, and inadequate attention to environmental and social issues. In developing the standard, the FSC drew on governmental and nongovernmental SFM principles, programs, and standards. These included the 1989 WWF certification proposal to the International Tropical Timber Organization (ITTO); the Rainforest Alliance's SmartWood certification program (established in 1989) and the Scientific Certification System's Forest Conservation Program (established in 1991); and the forest principles and indicator measures developed through various international processes (for example, the Rio Forest Principles, Agenda 21, and the Helsinki process). The FSC standard that emerged consisted of a set of international sustainable forest management principles and criteria that focused on addressing the protection of old-growth forests, prevention of illegal logging, protection of endangered species and habitat, restriction in the use of chemicals, enhancement of well-being of local communities, shared benefits from the forests, and respect for indigenous peoples' rights – all issues that had been inadequately addressed in the intergovernmental negotiations.[8] The FSC certification program also included an independent audit process and an eco-label.

The institutional strength of the FSC program as a voluntary, global multi-stakeholder initiative was unprecedented. Over 130 participants from twenty-four countries attended the 1993 founding meeting in Toronto. The FSC structure and process mimicked the constitutional norms and mechanisms of democratic states and created a new private political arena for sustainable forestry deliberation and global forest governance decision making (Tollefson, Gale, and Haley 2008). With the financial support of several European governments, but by leveraging global market forces rather than state-based authority, the FSC was able to steer past the impasse of the intergovernmental effort and succeed where states had failed – by reaching consensus on a set of global SFM rules that would be enforced through independent audit.[9]

In order to generate certification demand, ENGOs launched market campaigns targeting large forest products customers in the United Kingdom,

Germany, the Netherlands, Belgium, and the United States. Buyers were approached and advised that unless they stopped buying wood products from "endangered" forests and insisted that their wood product purchases were sourced from FSC-certified forests, their stores would be boycotted. In response, customers turned to their forest product suppliers and requested FSC certification; in some cases, they cut off demand for certain "high conservation value" forest products. For example, in early August 1998, the Rainforest Action Network (RAN), the Action Resource Center, and Earth First! hung a giant five-storey banner that said "Home Depot: Stop Selling Old Growth" at the corporate headquarters of the largest home improvement retailer in the world. Led by the Coastal Rainforest Coalition (subsequently renamed ForestEthics), protesters also mass-mailed postcards to the company, dressed up as bears and urged shoppers not to buy wood from endangered forests, sent a Great Bear Rainforest exhibit bus to a shareholders' meeting, and installed a Home Depot protest billboard over a clearcut forest area near Vancouver. Movie stars, sports heroes, and rock bands were asked to call for Home Depot to phase out the sale of old-growth wood, and a "kids campaign" mobilized 3,000 schoolchildren to send letters to the Home Depot CEO asking for a "gift of healthy forests." A year later, Home Depot responded by announcing that it would stop selling wood from endangered forests by the end of 2002 and establish a procurement preference for products certified as sustainable by the FSC.

The PEFC and National Forest Certification Programs

Although the FSC was intended to supplement state-based forest law and provide a market-based reward for progressive SFM efforts, most forest industry associations initially perceived it as a regulatory threat and many governments saw the institutional capacity of the FSC as a potential challenge to state sovereignty. In the mid to late 1990s, forest industry associations and governments around the world responded by developing their own competing forest certification programs[10] (Table 3.1). For example, the Canadian Pulp and Paper Association (CPPA) initiated a multi-stakeholder process under the Canadian Standards Association (CSA) to develop the CAN/CSA-Z809 forest certification standard (first published in 1996). In the US, the American Forest and Paper Association (AF&PA) revised its membership code of conduct to establish the Sustainable Forestry Initiative Principles and Implementation Guidelines (SFI) certification standard (1995). The Malaysian government established the Malaysian Timber Certification Council (MTCC) in 1998. Indonesia formed an Eco-labelling Institute in 1994 and established the Lembaga Ekolabel Indonesia (LEI) SFM standard in 1998. In 1999, European small private forestland owners established the Pan-European Forest Certification Programme (PEFC) (renamed the Programme for the Endorsement of Forest Certification in October 2003).

Table 3.1

Forest certification programs

Certification program	Initiated by	Year created
Forest Stewardship Council (FSC)	NGOs together with forest industries	1993
Sustainable Forestry Initiative (SFI)	American Forest and Paper Association	1995
Canadian Standards Association (CSA) SFM System	Forest Products Association of Canada (formerly the Canadian Pulp and Paper Association), together with the Canadian Standards Association	1996
American Tree Farm System SFM Standard (ATFS)	US non-industrial forest owners	1998
Lembaga Ekolabel Indonesia (LEI)	Indonesia Eco-labelling Institute	1998
Malaysian Timber Certification Council (MTCC)	Ministry of Primary Industries; Malaysian Timber Council	1998
Pan-European Forest Certification Programme, later renamed the Programme for the Endorsement of Forest Certification (PEFC)	Associations of small forest owners	1999
Sistema Chileno de Certificación Forestal (CERTFOR)	Chilean government; Chile's wood manufacturers' association	2002
Australian Forestry Standard (AFS)	Ministerial Council on Forestry, Fisheries and Aquaculture; forest industry	2003
Sistema Brasileiro de Certificação Florestal (CERFLOR)	Brazil Ministry of Development, Industry and Trade	2003

Source: Ozinga 2004, 34.

Other subsequent programs included the Chilean System for Sustainable Forest Management (CERTFOR), the Brazilian Forest Certification Program (CERFLOR), and the Australian Forestry Standard (AFS).

There are now over fifty voluntary forest certification standards globally, but four of them account for about 97 percent of the certified forests worldwide (see Appendix B). These include the FSC and three standards under the PEFC program: CAN/CSA-Z809, SFI, and the American Tree Farm System (ATFS) (for US small private landowners).

Table 3.2

Certified global forest area by major certification system, 2008

System	Area (millions of hectares)
Forest Stewardship Council (FSC)	103.5
Programme for the Endorsement of Forest Certification (PEFC)	202.3
Sustainable Forestry Initiative (SFI)	61.4
Canadian Standards Association (CSA)	75.8
American Tree Farm System (ATFS)	12.1
Sistema Chileno de Certificación Forestal (CERTFOR)	1.8
Sistema Brasileiro de Certificação Florestal (CERFLOR)	0.9
Australian Forestry Standard	7.9
Malaysian Timber Certification Council (MTTC)	4.7
Total FSC, PEFC, Malaysian	322.6

Sources: http://www.fsc.org; http://www.pefc.org; http://www.mtcc.com.my.

Increasingly, the portfolio of major certification schemes has been reduced to two lead systems (FSC and PEFC), as many national standards are gaining endorsement under the international PEFC umbrella program. For example, both the CAN/CSA-Z809 and SFI standards were endorsed under the PEFC in 2005, and the revised ATFS standard was endorsed in August 2008. The PEFC program now includes approximately thirty endorsed national standards and, as shown in Table 3.2, covers over two-thirds (roughly 68 percent) of the total global certified forest area, compared with the FSC's approximately 32 percent. In addition, some forests have been "dual-certified" – certifying with multiple certification schemes at the same time. For example, approximately 1.5 million hectares of Sweden's 6.9 million certified hectares are certified to both the FSC and PEFC standards.

Certification Effectiveness

Although certification is gaining acceptance and is clearly encouraging incremental forest management process improvements (such as better planning, consultation, inventory analysis, and so on), the forest health and productivity impacts of certification are largely unknown. Part of the difficulty is that connecting the indirect effects of process improvements (greater public consultation, stakeholder learning, and increased transparency, for example) to environmental outcomes is complex. Another challenge is timing. It is simply too early to measure some effects. For example, many forest improvements, such as biodiversity effects, will not be realized until well into the next harvest cycle, which in some forest communities could be as long as seventy to eighty years from now.[11] In addition, it can be difficult to disaggregate causation elements – that is, whether certification or some other internal or external factor caused the changes in the forest and forest management behaviour. Finally, the ultimate purpose of certification is a point of ongoing debate, thus leading to confusion over the definition and measurement of certification effectiveness. Is the intent to reward the good practices of the forest leaders or to gradually raise the performance of the laggards? For example, it is mainly the "good actors" in the North, with already well-documented forest plans and procedures, who have sought and readily achieved certification. In reinforcing and rewarding leading practices rather than necessarily improving lagging forest management, how effective is certification in resolving forest management problems?

Given these many challenges, some studies have looked beyond the direct "problem-solving" ability of certification to its indirect social, political, and economic consequences, such as creating a market advantage for large-scale forest operators in developed regions, and encouraging the development of similar private standards in other industry sectors (Auld, Gulbrandsen, and McDermott 2008). Some studies have also examined certification effectiveness by simply looking at uptake – who has certified and at what rate, within and across various regions?[12] By this measure, certification is achieving success in some regions (Europe and North America) while failing in others (tropical countries).

Certified forest area has increased sharply over the past decade to 320 million hectares, but this represents approximately only 8 percent of the 3.9 billion hectares of global forests. In addition, as has been mentioned, most of the certification has occurred in northern temperate regions. Although a key initial driver of the certification concept was concern about tropical forest degradation and deforestation, it has turned out that the certified areas are typically the better managed and expanding northern forests (see Figure 3.1). Less than 1 percent of southern tropical forests have been certified, and most is managed plantation forest, not natural rainforest. Factors contributing to the low uptake in tropical regions include a lack of demand from Southeast Asian customers for certified products; the lack of a price premium; complex, often poorly defined property rights; and, ultimately, high costs of implementation due to the lack of a strong regulatory framework (Cashore et al. 2006).

Certification adoption has been uneven not only between geographic regions but also between the various forest owner categories. A key barrier to successful participation has been that most programs are designed for larger industrial operators with the resources to develop forest management plans and a land base large enough to address biodiversity forest values. Certification among small landowners, particularly in developing regions, has been minimal. Although most certification programs have now developed

Figure 3.1

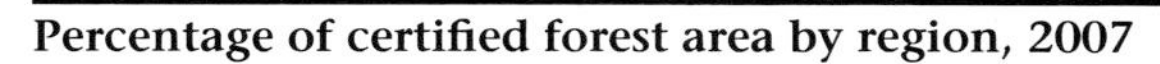

Percentage of certified forest area by region, 2007

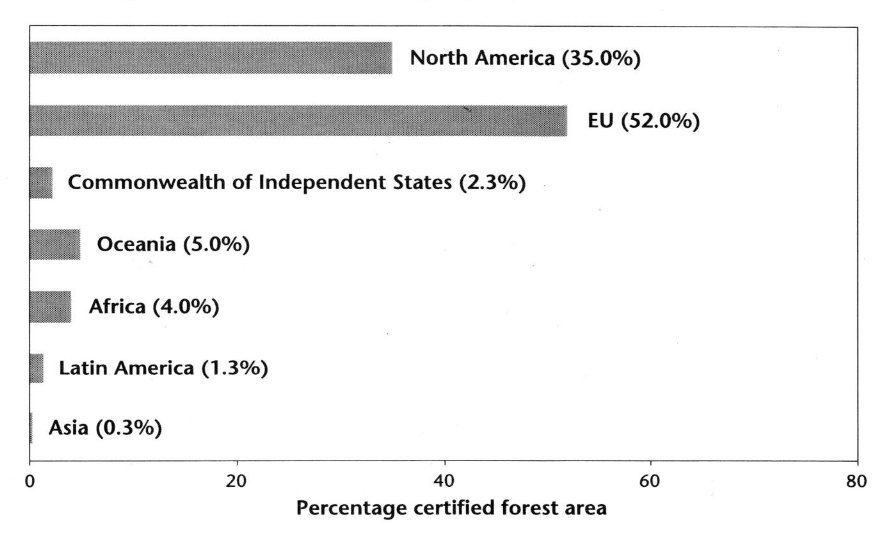

Source: UNECE/FAO 2007, 111.

options for the small private landowner (for example, group certifications), creating incentives to cover implementation costs and ensuring applicability of certification to this group of forest owners remains a challenge across all regions.

Certification Drivers

Certification is voluntary and adds costs, so in the absence of a regulatory penalty, why would a forest owner or operator participate? It might be expected that there is a market benefit, particularly a price premium for certified products, but this has not been the case. Most forest product customers have been unwilling to pay more – a premium is the exception rather than the rule. Other factors have therefore been driving certification uptake.[13]

Initially, industry associations were a key driver, requiring member companies to become certified as a condition of membership. For example, in January 2002, the Forest Products Association of Canada (FPAC) committed its membership to achieving SFM certification (CAN/CSA-Z809, SFI, or FSC) on all lands under their management by the end of 2006. One month before this commitment was made, there were 17 million hectares certified; three years later, the area had quintupled, to 86.5 million hectares. Similarly, the AF&PA initiated SFI participation by announcing in 1995 that enrollment in the program would be mandatory for all of its industry members.

Figure 3.2

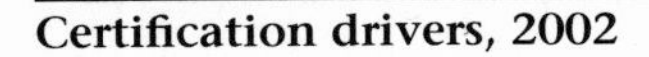

Certification drivers, 2002

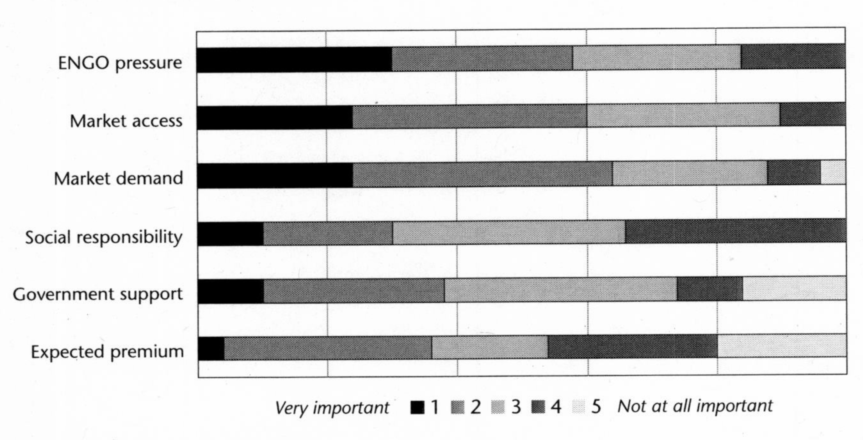

Source: UNECE/FAO 2002, 15.

As the greatest degree of certification adoption has been in jurisdictions with well-established and enforced forest management legal frameworks, fear of increasing government regulation has been a contributing driver. Companies have also certified their operations in response to advocacy pressures. Figure 3.2 shows the results of the UNECE Timber Committee's 2002 member survey, which highlighted ENGO pressures, customer demands, and market access as important certification drivers. At this time, forest companies were pursuing certification in order to meet anticipated growing export demands for certified forest products, with customer specifications being largely driven by pressure from ENGOs. Certification was voluntary but companies were afraid of losing access to offshore customers if their forests and forest products were not certified. Although certification markets are slow in developing, maintaining market access in the face of ENGO boycott pressures has been and continues to be a significant certification driver.[14] Also, despite the lack of a consistent or sufficient price premium, many companies identified a business case for certification in terms of enhanced corporate reputation, supply chain efficiencies, mitigation of risk, and achievement of continual improvement in forest ecosystem conditions (Metafore 2004, 5-6). As well, the business case for certification is increasing with climate change initiatives that include the development of carbon markets and increased demand for biofuel.[15]

In some cases, companies have pursued forest certification out of commitment to social responsibility and an "enlightened self-interest" in being good corporate citizens and maintaining their "social licence to operate."

Increasingly, however, they are approaching certification as simply a necessary cost of doing business rather than as a means of demonstrating CSR and/or achieving a particular competitive advantage. For example, the vast majority of certified forest and paper products continue to be marketed without any reference to certification.

A fundamental determinant of certification uptake is whether and on what basis forest owners consider certification to be a legitimate form of forest rule. Drawing on sociological theory, Cashore argues that forest companies have granted certification various forms of pragmatic, moral, and cognitive legitimacy (Cashore 2002).[16] Pragmatic legitimacy concerns the instrumental or business case for certifying. Companies that certify because it is "the right thing to do" have granted certification moral legitimacy. With cognitive legitimacy, companies participate in certification as it is deemed "an accepted cost of doing business." Although the pragmatic business case initially dominated and continues to influence companies to pursue certification, there appears to be a gradual evolution towards the inclusion of moral and cognitive drivers. Drawing on neo-institutional theory, Bernstein and Cashore describe this as a shift from a "logic of consequence" based on pragmatic individual rational calculation to a "logic of appropriateness" that combines moral and cognitive legitimacy (Bernstein and Cashore 2007, 355). They argue that although not yet empirically tested, this is a positive shift, as institutions that are granted moral and cognitive legitimacy tend towards greater durability than those based on pragmatic cost/benefit calculation. In other words, moral and cognitive legitimacy increases the likelihood that certification systems will persist.

State engagement can bestow moral and cognitive legitimacy on certification, thereby enhancing its institutional durability. As the forest management legislator, government has the potential to send a very strong signal that certification is an accepted forest rule and expected forest activity. Government provision of certification information, technical guidance, and/or financial incentives; establishment of procurement policies; certification of public land; and/or legislation of certification uptake have all been important certification drivers.

Competition between the FSC and the various national certification programs has also contributed to certification uptake. For example, industry associations in Canada and the US encouraged certification largely as a defensive tactic to secure support for competitor national certification schemes rather than the FSC certification program, in order to prevent regional marginalization. The PEFC and FSC continue to adjust their standards relative to each other in order to gain legitimacy and facilitate adoption.

To summarize, this analysis of forest certification drivers draws attention to three fundamental points. First, certification appears to be developing into a more durable institution as the motivation for its adoption evolves from mainly pragmatic, cost/benefit reasons towards inclusion of cognitive and moral considerations that certification is "simply the right thing to do." Second, although forest certification is described as a market-driven mechanism, social and governmental drivers are also influencing uptake. Finally, the competition between the FSC and the various national certification programs encourages ongoing revisions to the standards that facilitate certification adoption.

Credibility and Mutual Recognition

There have been and continue to be philosophical differences, extensive debates, and considerable confusion about the merits and drawbacks of the various forest certification programs.[17] ENGOs assert that the FSC is the only credible standard. Small forestland owners generally participate in PEFC programs such as the American Tree Farm System. Industry supports the availability of competing standards, as well as "mutual recognition" between the FSC and PEFC programs. Governments also support choice and mutual recognition.

Two key points of debate have concerned: (1) the degree to which the standards prescribe performance requirements, and (2) the extent to which NGOs or industry influence the standard development process. In general, industry has considered the FSC an ENGO-driven standard with overly prescriptive performance requirements. ENGOs criticize the PEFC (including the CAN/CSA-Z809 and SFI standards) for being "industry tick-box exercises that exclude social and environmental NGOs" and for failing to adequately address key sustainable forestry issues such as plantations, old growth, and indigenous peoples.[18]

Groups such as Greenpeace, the WWF, and Friends of the Earth refuse to accept mutual recognition as they do not view the PEFC and FSC standards as equivalent and fear that mutual recognition would put downward pressure on the FSC requirements.[19] Industry and governments are supportive, however (Griffiths 2001). They argue that it is important to maintain flexibility of choice between certification programs as one standard will not easily address the diversity of forest types and ecosystems or the wide range of forest tenures and operating arrangements, and a lack of choice will potentially create market distortions. In some instances, political tensions between groups are heightening, as in the case of the US Green Building Council's Leadership in Energy and Environmental Design (LEED) program decision to recognize only the FSC standard in its certified wood credit criteria.

Despite the growing acrimony and impasse in some sectors, discussions about mutual recognition continue and now focus largely on legality and sustainability definitions in the context of large retailers' implementation of green procurement policies and countries' establishment of trade policies to avoid importation of illegal timber.

In order to alleviate confusion and avoid uninformed, biased preference for one scheme or another, governments, businesses, and NGOs have developed various criteria to assess whether a certification program is credible. These include:

- Openness – Does the certification system provide opportunities for input and participation by stakeholders?
- Transparency – Is the certification decision-making process conducted in a way that is visible and transparent to interested parties?
- Freedom from bias – Does the certification decision-making body include an array of interests and backgrounds?

In addition, credible certification schemes include an accreditation process to ensure the capability of certification auditors, and accepted certification standards have a requirement for third-party independent certification audit and regular independent monitoring.

Although the PEFC and FSC standards incorporate all of the foregoing credibility criteria,[20] groups with a political agenda to promote or denigrate a particular standard continue to introduce additional attributes that are referred to as "legitimacy threshold criteria."[21] These new criteria both fuel the debate and spawn new considerations. Fundamentally, the competition between certification standards constitutes an adaptive process that is providing critical feedback and encouraging ongoing changes to both the PEFC and FSC standards.[22] Rather than being driven further apart, the PEFC and FSC standards have become increasingly similar in design and SFM content as they compete for acceptance.

Through several rounds of revisions, all of the leading certification programs now include performance and management system elements, multistakeholder oversight, third-party audit requirements, and eco-label and chain-of-custody certification options. There is already mutual recognition *within* the two leading certification schemes, as the FSC and PEFC have together accredited or endorsed over fifty national or subnational standards. Increasingly, mutual recognition is occurring *between* the FSC and PEFC as customers, governments, financial institutions, and industry associations request that forests be certified to one of the accepted SFM certification programs. Large global forest and paper customers such as Centex Homes, Hallmark Cards, Lowe's, Office Depot, Staples, and Time Inc. all have inclusive purchasing policies that recognize the various SFM certification standards. Even financial institutions such as the $240 billion CalPERS Fund have established sustainable forest management standards stating that their timber investments require certification of the forestland "by an independent third party." As well, public and private landowners are mutually recognizing the various standards by dual-certifying forestlands to both FSC and PEFC standards, and certification audit organizations are now offering dual-audit services. Finally, some governments are mandating public land certification to at least one of the SFM systems, and are recognizing both the PEFC and FSC programs by establishing inclusive public procurement policies that accept certification to any of the recognized credible certification standards. While appreciating the political tensions and underlying differences between the standards, a comprehensive analysis of forest certification as a governance mechanism therefore requires an inclusive approach that considers the overall dynamic of both PEFC and FSC rule-making systems.

Returning to the earlier discussion, forest certification emerged because it was needed: national governments were failing to adequately address the rapid global loss of tropical forests. Non-state actors established the FSC as a private mechanism to fill this international forest governance gap. Rather than spurring improved governance of tropical forests, however, the FSC sparked the development of competing national certification standards and the certification of northern temperate and boreal forests, with resulting overlap of certification systems with public forest law in industrialized timber-producing countries. As argued in the next section, certification is still needed, but the nature, dynamics, and purpose of certification have changed. In the high public capacity regions where certification is operating, it is not filling a governance gap so much as serving as an additional policy instrument that is helping to regulate forest activities through hybrid co-regulatory governance arrangements.

Classifying Forest Certification Governance

Forest certification is a new governance phenomenon. Through multi-stakeholder decision-making bodies that closely resemble traditional democratic institutions, corporations and NGOs develop, implement, and enforce transnational private forest rules by leveraging market supply chains. Forest certification is an example of voluntary self-regulation, corporate social responsibility, and a private environmental governance mechanism, but it is also distinct within each of these classifications (see Table 3.3). It is important to understand these differences, as forest certification is in many respects an *ideal* form of private governance compared with other voluntary initiatives. As well, these distinctions are a source of governance strength.

Comparing forest certification with other examples of voluntary initiatives reveals certain distinctions, noted in Table 3.3. For example, in contrast to self-regulatory codes of conduct such as the chemical industry's Responsible Care program, certification systems are open, transparent, multi-stakeholder,

Table 3.3

Forest certification private governance classification

Private governance classification	Description	Forest certification private governance distinction
Voluntary self-regulation	Voluntary rules and standards are developed and enforced by private organizations.	Forest certification *also* includes multi-stakeholder decision making and independent enforcement (unlike an industry association code of conduct).
Private environmental governance mechanism	Private actors are involved in authoritative rule making that was previously the prerogative of governments.	Forest certification gains rule-making authority through market acceptance, not government delegation.
Corporate social responsibility	Corporate voluntary initiatives integrate social and environmental stakeholder expectations (often going beyond legal requirements).	Forest certification systems *also* include a multi-stakeholder rule-making and adjudication body, a prescriptive standard, and an independent enforcement mechanism.

and independently audited, as opposed to closed, industry-dominated, and driven by association membership.[23] In contrast to eco-labelling programs (such as the German Blue Angel and the Canadian EcoLogo programs), certification systems have dynamic, continually improving requirements rather than static environmental quality measures. And, finally, compared with other CSR standards (ISO 14000, the Global Reporting Initiative, the United Nations Global Compact, and so on), certification systems prescribe "hard rules" rather than flexible commitments, turn to the market and global supply chains to create incentives, and include third-party audit enforcement mechanisms.[24]

Non-State Market-Driven (NSMD) Governance

Environmental governance scholars have adopted various labels to capture forest certification's unique governance qualities, including "civil regulation," "transnational business regulation," "supra-governmental regulation," "private hard law regulation," and "non-state market-driven governance" (NSMD).[25] All of these terms highlight the private regulatory capability of forest certification systems. As Cashore and colleagues (2007) explain, beyond leveraging global supply chains and including a third-party independent audit enforcement mechanism, NSMD systems are distinct private governance mechanisms as they establish dynamic deliberative forums (not dominated by business interests) and they develop prescriptive hard law rules.[26] Ultimately, these characteristics contribute governance capacity that provides NSMD systems with private rule-making authority beyond that of other self-regulatory CSR initiatives.

As well, a key source of NSMD systems' distinctiveness is the location and source of governance authority and the role of government. As summarized in Table 3.4, the location of policy authority is in market transactions, the source of authority is evaluation by external audiences, and governments do not play the role of policy maker but rather are considered an interested group, with indirect facilitating or debilitating influence. The theory further elaborates that if and when governments use their sovereign authority to require adherence to private standards, then the concept of NSMD governance will cease to exist, as the system will no longer be market-driven but rather government-driven. The absence of public authority is a categorical condition of NSMD governance. Although government's role in certification is acknowledged as important, NSMD theory limits state engagement to actions that are deemed as not invoking government's sovereign authority. As Cashore (2002, 510) outlines, these include:

- acting through existing policy rules that play a background role in NSMD systems
- acting as a traditional interest group in the NSMD policy-making process
- initiating procurement policies
- acting as a landowner certifying their own public lands
- providing resources to help groups certify
- providing expertise in the development of standards.

Under the NSMD classification, forest certification is considered a purely private authority mechanism, establishing private forest rules independent of state authority – a critical assumption that this book challenges.

Forest Certification in International and Domestic Rule-Making Contexts

Forest certification operates as a form of private environmental regulation alongside state policy and laws at both the international and domestic levels. In the global political arena, forest certification overlaps with multilateral institutions and international state-led cooperative forest governance agreements, laws, and conventions. At the domestic level, certification replicates the law making functions of governments. In each instance, certification private authority serves to complement and supplement the existing regimes rather than supplanting traditional state authority.

Table 3.4

Authority in non-state market-driven (NSMD) governance compared with other forms of governance

Features	Traditional government	Shared governance	NSMD governance
Location of authority	Government	Government gives ultimate authority (explicitly or implicitly)	Market transactions
Source of authority	Government's monopoly on legitimate use of force; social contract	Same as traditional government	Evaluations by external audiences, including those it seeks to regulate
Role of government	Has policy-making authority	Shares policy-making authority	Acts as an interested group, landowner (indirectly facilitates or debilitates)

Source: Cashore 2002, 504.

Certification as a Global Governance Mechanism

In terms of global forest governance, forest certification behaves as transnational private regulation. Certification systems have gained legitimacy and institutional strength to establish and promote rules and norms in the issue area of global sustainable forest management. PEFC and FSC certification programs are therefore contributing to the overall global forest regime, operating alongside multilateral hard law conventions (the Convention on Biological Diversity, the Convention on International Trade in Endangered Species of Wild Fauna and Flora, and others) and soft law agreements (such as the non-binding UN Forest Principles).

The public and private mechanisms within the global forest regime have strengths and weaknesses. In general, state-based multilateral mechanisms are slow and encumbered by national interests, but are generally structured to encourage broad inclusion and are accountable and enforceable through international organizations and state legal authority. Private certification systems are flexible and responsive and transcend borders to address local to global concerns, but they also lack democratic accountability and, as a voluntary mechanism, have weak enforceability and uncertain durability.[27] Neither certification nor intergovernmental processes have achieved solutions on their own; together, however, they offer the potential for more effective global forest governance. Ideally, public and private mechanisms interact with each other in the global arena, compensating for each other's weaknesses and drawing on their respective strengths to contribute to greater effectiveness of global governance.

Conceptually, there is a difference of perspective among scholars of global forest politics as to whether private forest governance authority can complement traditional state-based international cooperative efforts within the forest regime. On the one hand, as Chapter 2 outlined, traditional statists view sovereignty as autonomous and authority as zero-sum, and therefore argue that private governance through market-based certification inherently competes with and potentially supplants traditional government authority, resulting in a retreat of state leadership in global cooperative efforts. Because forest certification is viewed as being in direct competition with state sovereignty, greater state intervention is called for to re-embed market forces within authoritative state control (Humphreys 2006; Lipschutz and Fogel 2002).

On the other hand, new governance scholars argue that post-sovereign arrangements have emerged in which private and public authorities operate in parallel within a broader interdependent hybrid political arena, thus supplementing overall global governance capacity and transforming rather than supplanting state sovereignty (Falkner 2003; Karkkainen 2004). Some describe this as a "new medievalism" in which governance authority is located simultaneously at multiple overlapping sites (Kobrin 1998). Thus, from the new governance perspective, forest certification can be complementary to traditional international mechanisms rather than an assumed competitive threat to state-based cooperative efforts (Gulbrandsen 2004; Haufler 2003; Pattberg 2006).

This book supports the new governance position regarding shared authority. If certification were supplanting public authority, there would be less participation by and support from international forest governance institutions, and greater state reliance on forest certification systems. This is not the case, however. Intergovernmental forest governance negotiations continue,[28] and governments that are engaging in certification are enabling certification as one tool in their traditional array of governing instruments. As argued in Chapter 2, new governance mechanisms such as certification, eco-labelling, and voluntary industry codes and agreements can complement rather than necessarily replace traditional regulation in co-regulatory arrangements.

Fundamentally, the state has not been in retreat with the emergence of certification. Rather than substituting for international institutions, certification is coexisting with state-based cooperative efforts, contributing to a multicentric global forest regime. Not only do certification programs overlap with international mechanisms by incorporating international SFM principles but, as Gulbrandsen (2004, 83) notes, there are also various dimensions whereby certification supplements rather than supplants international forest governance, including:

- enabling greater inclusion and balancing of powers of economic, ecological, and social groups in forest governance decisions
- ratcheting and/or strengthening of environmental and social performance standards
- ensuring enforcement of sustainable forest principles and encouraging the continual improvement of forest practices through third-party, independent auditing
- engaging a large spectrum of forest producers across a wide geographic reach
- promoting trade in sustainable forest products through chain-of-custody certification and eco-labelling.

Thus, given the respective strengths and weaknesses of international and forest certification mechanisms, their interaction can be mutually beneficial. This occurs within a flexible, multi-tiered global forest regime with multiple overlapping sites of public and private agenda-setting and decision-making authority.

While debate continues regarding the implications of private governance authority in the global arena, the more complex dynamic is actually playing out at the domestic level. In the international realm, where there is no sovereign world government, it can be argued that certification is filling a governance gap. At the domestic level within developed countries, where certification is largely occurring, not only is there an established sovereign state authority but there are also already well-established forest institutions. With forest rules and forestry agencies firmly in place, public and private rule-making authorities are directly overlapping within these high public capacity regions. The result is hybrid co-regulatory forest governance arrangements that go beyond the theoretical NSMD classification.

Forest Certification as Domestic Forest Law

Within the domestic policy environment, certification replicates the agenda-setting and forest policy implementation and enforcement functions of governments. Certification is essentially a nongovernmental law-making mechanism (Meidinger 1997, 2003a; Tollefson, Gale, and Haley 2008). As explained earlier in this chapter, certification is classified in the environmental governance literature as a purely private, non-state market-driven governance mechanism and private hard law regulation. Employing the co-regulatory map of policy instruments introduced in Figure 2.3, this situates certification in the bottom left quadrant, constituting a market-driven self-regulatory policy tool with non-delegated private governance authority (see Figure 3.3). As shown, forest certification can be considered a distinct policy mechanism, in contrast to traditional state-based policy instruments such as command-and-control regulation and delegated self-regulation, which reside within traditional state authority.

Figure 3.3

Forest certification as a non-state market-driven policy instrument

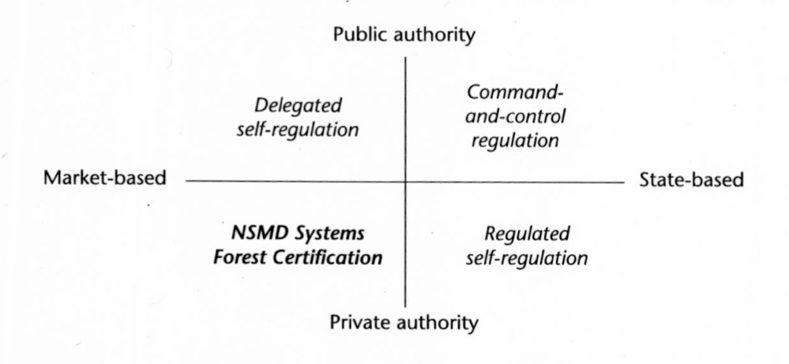

Certification *is* a distinctive policy instrument in terms of its initiation by private actors and its non-delegated private governance authority outside of formal state sanction. This, however, is only a partial account of the forest certification governance dynamic. Upon closer examination, forest certification is not entirely market-driven, nor do certification systems constitute purely private authority. For example, in their formulation, certification rules overlap with and depend on existing forest laws and legal frameworks. Implementation hinges on policy alignment and regulatory compliance, and governments play a key role in overseeing, facilitating, legitimating, and even enforcing certification standards. In other words, certification constitutes a co-regulatory governance mechanism with coincident public and private rule-making authority that is "driven" by governments as well as markets and civil society.

Fundamentally, although the NSMD classification treats the state as an interested party on the same level as other landowners and participating groups, governments are different. In their inherent capacity as sovereign law maker, governments have special status and are therefore unlike any other stakeholder. Thus, any government certification activity will have implications that transcend the influence of other groups. As shown in Table 3.5, many of the government activities that NSMD theory lists as "acceptable" because they do not invoke state authority (such as initiating procurement policies, creating financial incentives, providing background legal framework, and so on) in fact have significant influence on certification design, implementation, and adoption precisely because they *do* engage state authority (shown by check marks). As well, several of the examples in the table (under "Overlapping governance") highlight how governments

Table 3.5

Overlapping public/private governance authority in forest certification

Governance function	Non-state authority	Overlapping governance	State also a driver	State authority engaged
Formulation (agenda setting)	Private actors initiate non-delegated self-regulatory standards.	Legislation encourages the initiation of self-regulation.	✓	✓
		Governments support the initiation of national certification schemes.	✓	
Negotiation (rule making)	NGOs and corporations deliberate over rules and process. Governments are refused formal participation.	Certification draws on forest policy to establish private rules.		✓
		Certification standards incorporate legal compliance.		✓
		Governments provide technical, administrative, and/or financial support for standard development.	✓	
Implementation	Private landowners implement certification.	Governments provide incentives to facilitate certification.	✓	
		Governments legitimize certification by certifying public land.	✓	
Monitoring and enforcement (compliance)	Third-party audits are performed by independent certification bodies.	Governments leverage certification audits to streamline regulatory compliance audits and monitoring.	✓	
	Private enforcement occurs through threat of loss of certificate if non-compliant.	Governments mandate forest certification on public land.	✓	✓

Figure 3.4

Forest certification co-regulatory governance

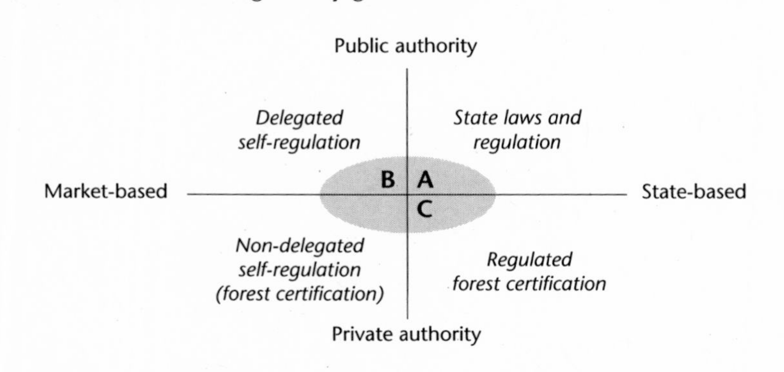

are driving certification acceptance and adoption by enabling and endorsing certification elements that overlap with state policy and functions (also shown by check marks).

Ultimately, NSMD classification is only a partial account of the forest certification governance phenomenon. Voluntary CSR standards such as forest certification *are* distinct with respect to their non-delegated private authority, but they also overlap with public authority and rely on public governance capacity. As I discuss next, certification can be more comprehensively understood as a co-regulatory governance mechanism with public and private rules and authority interacting at the formulation, implementation, and enforcement stages.

Forest Certification Co-Regulatory Governance

Although certification and government programs overlap and interact, there has been surprisingly very little investigation of the role of government in forest certification or of the interplay between forest certification and forest policy.[29] As mentioned previously, research for the most part has conceptualized forest certification as a purely market-based mechanism independent of state authority. Another reason for the lack of inquiry into certification co-regulatory governance is that forest certification systems are still a relatively new phenomenon, barely a decade old in most jurisdictions. The public/private dynamic has only recently begun to play out as adoption of certification has increased and as the standards and state forest regulations go through their respective revision cycles. Understanding and managing this interplay therefore constitutes an emerging area of policy research as well as an applied governance challenge. Employing the governance matrix

Figure 3.5

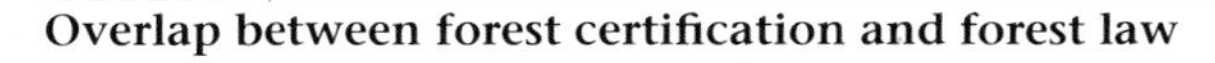

Overlap between forest certification and forest law

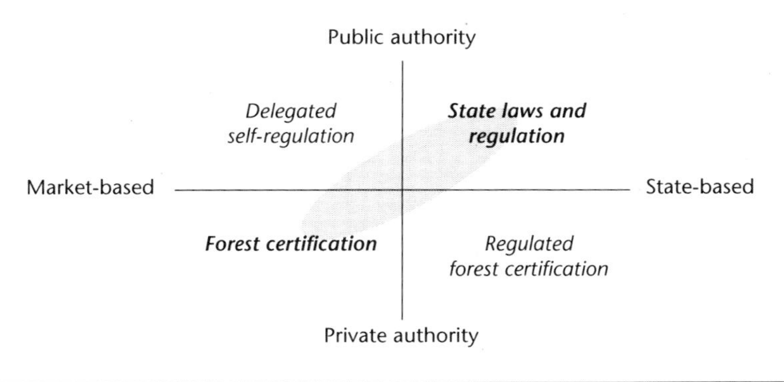

developed in Chapter 2 (Figure 2.3) reveals how forest certification overlaps and interacts with traditional regulatory approaches within a co-regulatory forest governance system. As shown in Figure 3.4, certification systems intersect with traditional forest law and state-delegated self-regulatory policy mechanisms (Areas A and B). In some instances, certification has also become directly embedded in forest legislation as an example of regulated self-regulation (Area C). These instances of certification co-regulation are outlined below.

Certification and Forest Law

Forest certification rules mimic, overlap with, and can go beyond public forest laws and regulation (see Figure 3.5). For example, in their comparative study of forest policy attributes across regions, McDermott and colleagues (2008, 67-68) conclude that "there is significant cross-fertilization between certification standards and government policies ... certification standards are largely shaped by state-based regulatory norms." Not only do state laws have an impact on the formulation, implementation, and enforcement of certification systems but forest certification also has an impact on forest laws and state forest administration. There is a dynamic exchange between the public and private rule-making systems.

For instance, in terms of the influence of state laws, certification relies on an existing regulatory framework to provide contract and property law as well as to enable the chartering of the certification body and the awarding of conformance certificates to specific forest owners and defined forest areas. Certification must also respect established forest laws and agreements in order to be considered legitimate and not be prohibited by governments.

Figure 3.6

Overlap between forest certification and self-regulation

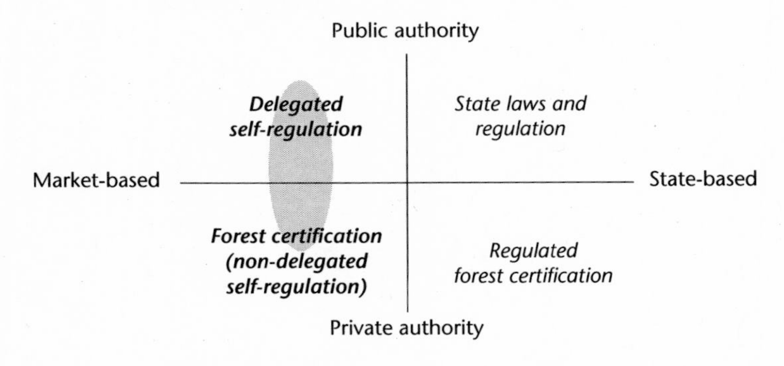

In their formulation, forest certification programs have therefore been designed to align with state-based SFM principles and criteria, as well as to include consideration and respect for local forest laws and international agreements.

Certification systems not only rely on public law to establish contracts between the various parties but can also establish formal partnership arrangements with governmental agencies to perform accreditation services in order to ensure the independence and competence of certification audit organizations. For example, the Standards Council of Canada and the Swedish Board for Accreditation and Conformity Assessment (SWEDAC), both of which are public agencies, accredit the PEFC certification auditors in Canada and Sweden.

Although certification is classified as an NSMD system and therefore considered an example of "governance without government," forest certification systems are, in fact, closely connected with and even influence public forest policy. State forest policies and rule-making processes play an important role in certification development, implementation, and enforcement. Ultimately, forest certification is a case of governance *with* government.

Certification and Self-Regulation

As outlined earlier in this chapter, forest certification systems are an example of voluntary self-regulation yet have certain distinguishing features. The main difference is that unlike traditional forms of self-regulation, such as professional codes, forest certification gains its legitimacy and rule-making authority through acceptance by private actors along the supply chain rather than from government alone. In other words, certification is an example of *non-delegated* self-regulation – the state has not officially handed over forest rule-making responsibility to private actors.

While distinct from *delegated* self-regulation, forest certification programs also share features, such as voluntary implementation and compliance, that overlap with state-sanctioned self-regulatory programs (Figure 3.6). Applying the range of self-regulation definitions presented in Chapter 2, Table 3.6 summarizes how certification can be classified under not one but rather various regulatory labels. Each of these classifications is described below.

Regulatory Scope

Forest certification schemes can be classified as both *group self-regulation* and *individual self-regulation*. For example, some standards (such as CAN/CSA-Z809 and SFI) develop firm-specific local rules, and all standards require conformity with an industry-wide set of principles and criteria. Certification can also be considered both an *economic self-regulatory* and *social self-regulatory* governance mechanism. Certification aims to correct for unaccounted negative environmental externalities such as deforestation, riparian damage, and so on by setting specific conditions for market entry, and also to address the protection of non-timber public goods such as cultural forest values and the maintenance of biodiversity by instituting "predictable, long term ordering of the behaviour of forestry firms" (Haufler 2001; Meidinger 2003a, 267).

Rule Making

In terms of private rule-making authority, when first introduced, certification schemes represented both *industry self-regulation* (for example, the SFI program in the US) and *multi-stakeholder self-regulation* (for example, the FSC and CSA programs). *All* of the leading programs are now pluralistic in their design, incorporating economic, environmental, and social stakeholders (including governments in some cases) in the negotiation, development, and revisions of the standards. Certification is an example of *pure self-regulation,* as rules are developed without formal state sanction.

Implementation

At the rule-delivery (implementation) stage, and in terms of Knill and Lehmkuhl's regulatory typology (see Chapter 2), forest certification is an example of *private self-regulation.* Private actors are responsible for implementation and ensuring conformity with forest certification SFM objectives independent of government intervention. Forest certification schemes can also be classified as *regulated self-regulation,* however. As outlined in the previous section, the standards incorporate and rely on legal compliance, and government agencies can oversee auditor accreditation.

Table 3.6

Overlap between forest certification and self-regulatory governance

Governance characteristic	Self-regulation mode	Overlap between forest certification and self-regulation
Regulatory scope	Group	There is conformity to industry-wide set of principles.
	Individual	Firm-specific local rules are developed under some of the standards (e.g., CAN/CSA-Z809).
	Economic	Corrects for unaccounted externalities (e.g., deforestation, riparian damage, etc.) by setting conditions for market entry.
	Social	Non-timber values such as cultural forest values and maintenance of biodiversity are protected through predictable, long-term ordering of the behaviour of forestry firms.
Rule making	Industry Multi-stakeholder	Some certification standards are developed solely by industry, but all standards now incorporate multi-stakeholder input.
	Pure	Corporate actors formulate the rules.
Implementation	Private	Private actors (industry and civil society) are responsible for implementation and ensuring conformity.
	Regulated	Implementation relies on regulatory compliance and government cooperation.
Enforcement	Voluntary	Certification rules are enforced through private, independent, accredited audit organizations.
	Mandated	Some governments are mandating forest certification on public land and including certification in public procurement policies.

Note: Self-regulation classifications are those employed in Haufler 2003; Knill and Lehmkuhl 2002; and Gunningham and Rees 1997, as explained in Chapter 2.

Enforcement

At the enforcement stage, and with respect to the Gunningham and Rees regulatory typology (see Chapter 2), forest certification is an example of *voluntary self-regulation.* Rules are formulated by industry, NGOs, and other non-state actors and enforced through independent, accredited third-party auditors. In recognition of an opportunity to brand their forest policy with a third-party stamp of approval, however, an increasing number of governments are formally intervening to mandate forest certification. This represents a shift from *voluntary self- regulation* to *mandated self-regulation,* as will be explained below.

If the above assessment has seemed at all confusing, it is because it *is* confusing. Scholars are applying different terminology (highlighted in italics) and different regulatory typologies to describe various forms of self-regulation at the various stages of governance. The complexity is compounded by the fact that forest certification does not easily conform to a single definitional category. Forest certification standards have differed in their formulation, design and evolution, and reliance on public institutions, thus spanning the various self-regulation classifications. What *is* clear from this analysis is that certification extends beyond its classification as a non-state market-driven system, overlapping with delegated forms of self-regulation – regulation that falls under the shadow hierarchy of the state.

Forest Certification and Regulated Self-Regulation

In industrialized regions (where certification is largely occurring), governments are increasingly adopting umbrella meta-governance approaches in response to voluntary certification, that is, they are regulating the self-regulatory forest certification mechanism. With government enrollment, certification becomes both a market- and a state-driven hybrid governance instrument, as shown by the shaded area in Figure 3.7.

For example, some governments are mandating forest certification, directly integrating certification into their statutory regimes. In addition, an increasing number of governments are incorporating certification directly into public procurement policies (McCrudden 2007; Simula 2006). Both of these are examples of regulated self-regulation, as the state is officially sanctioning a private self-regulatory initiative by requiring compliance. Governments are also directly integrating certification into the public administration of forestlands by certifying state-owned forests. They are also directly leveraging certification to increase the overall resources of the state, for example, by using certification audits to supplement state monitoring programs and forest law enforcement.

Thus, in advanced industrialized countries with well-maintained legal systems and a strong regulatory enforcement presence, it would appear that

Figure 3.7

Forest certification and regulated self-regulation

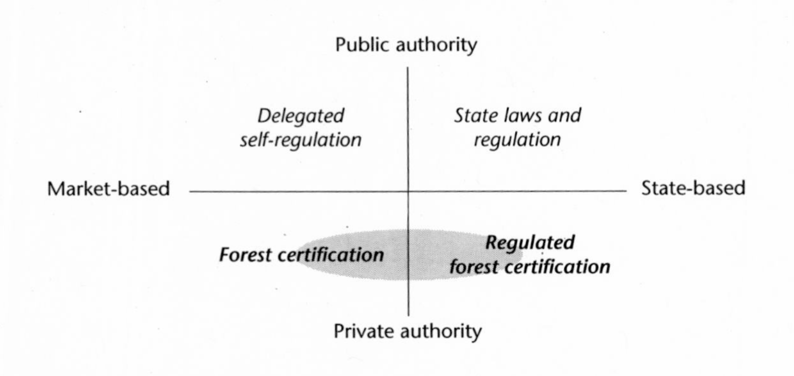

governments are not retreating in response to certification so much as redefining their regulatory roles towards co-regulatory approaches that leverage and facilitate CSR initiative. These examples are illustrated in the cases presented in Chapters 4, 5, and 6.

The Spectrum of Government Role in Forest Certification

Forest certification is potentially a positive development for governments simply because the private sector is voluntarily taking on traditional public responsibilities to ensure the sustainable management of the forest resource. Presumably, therefore, it should save governments time and money. With increasing adoption of certification, the state ideally moves towards a co-regulatory role, enabling and leveraging private governance authority as appropriate. Still, the public/private balance and forest governance outcomes are largely determined by the baseline laws and administrative functioning of the state, and how government positions itself in response to private certification initiatives.

Inherently, governments are key actors in certification as they regulate forest practices, own public forestland, and are significant buyers of forest and paper products. In the large *consumer* countries, the key government role in certification is the establishment of procurement and trade policies that favour certified forest products. Within timber-*producing* nations, governments play a significant role in certification by establishing forest laws and providing a supporting regulatory framework. Because certification systems overlap with state forest law, governments also ensure the ongoing alignment and congruity of certification programs with the state's forest management goals and objectives. Ultimately, the challenge for governments lies in determining their optimal response to certification so as to leverage the adaptive, flexible, and innovative properties of private rule-making mechanisms within the domestic forest policy mix, while maintaining policy sovereignty to regulate SFM accountability.

Rationale and Benefits of Government Engagement

Fundamentally, the rationale for government engagement in forest certification is to ensure fair market play (economic rationale) and a desirable quality of sustainable forest management (ecological and social rationale) (Rametsteiner 2000). Governments benefit in several ways from supporting certification: potentially enhanced stakeholder agreement on SFM, improvements in forest management, possible reduction in enforcement and monitoring costs, and greater market and public confidence in forest policy. Some governments have encouraged certification as a means of branding their local forest practices as superior and of promoting their forest industry's competitiveness in global markets. As well, governments that lack financial and human capital and technical resources have an incentive to leverage certification to achieve forest policy objectives.

In October 2005, the United Nations Economic Commission for Europe's (UNECE) Timber Committee held a policy forum on "The Role of Government in Forest Certification." Government, industry, and NGO participants from developed and developing countries identified a range of government roles (Koleva 2006). The five common expectations of government included:

- ensuring compliance of SFM standards with laws and regulations
- intervening in certification to prevent monopolies, unbalanced market conditions, and/or trade distortions
- participating as a neutral party in moderating between the different certification programs and encouraging mutual recognition
- preparing public procurement rules that are inclusive, non-discriminatory, and harmonized between countries
- offering certification technical and financial assistance for capacity building in developing regions.

Government Concerns with Certification

A key issue that has prompted all states to keep a close eye on forest certification is the potential for certification to act as a technical barrier to trade and create competitive disadvantage.[30] Governments are also concerned about the implications of increasing private authority for democratic processes and state sovereignty over the domestic forest policy agenda.

Figure 3.8

Median average total costs for certification in Canada and the US by ownership size, 2007

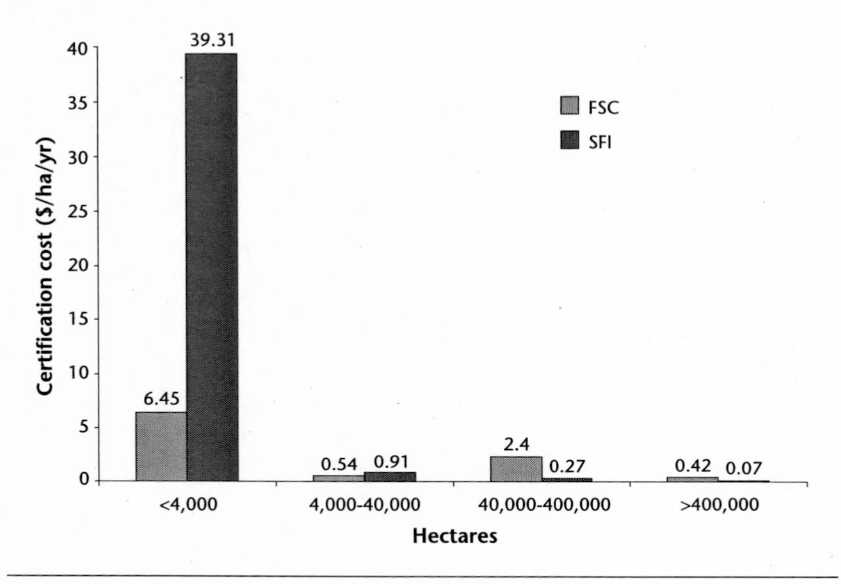

Source: Cubbage and Moore 2008; Cubbage et al. 2008.

Government response to certification has therefore included consideration of the intended and unintended effects of certification on forest owners, the forest economy, and forest policy objectives. In particular, as outlined below, governments have focused on ensuring that certification does not discriminate against small private forest owners and operators, create trade distortions, and/or introduce disincentives to sustainable forest management (Rametsteiner 2002).

Discrimination against Small Landowners

Certification increases the overall financial costs of forest management, which translates to a larger per-hectare certification expense for the small forestland owner than for the larger industrial operator.[31] For example, as shown in Figure 3.8, there is a significant difference in the median average total costs per hectare per year of the FSC (in the US) and SFI (in Canada and the US) for certified forests under 4,000 hectares compared with the larger certified forest areas, with costs ranging from $0.07 per hectare to over $40 per hectare. In response to the cost inequity, some governments have provided smaller forest owners with informational and technical support, as well as financial incentives to minimize potential market discrimination.

Trade Distortion

Forest concession operators in tropical regions are mainly small to medium-sized enterprises. These regions typically have weak regulatory frameworks and low forest law implementation and enforcement. Small forest operators generally lack forest management plans and documented procedures and, in the absence of a price premium for certified timber, cannot bear the increased marginal costs associated with certification.[32] Thus, for developing regions, market requirements for certified timber can act as a non-tariff barrier to trade, essentially barring access to premium markets such as Europe, Japan, and North America and resulting in the diversion of their forest products to regions with less discerning legality and sustainability requirements (Egypt, China, India, and so on).

An underlying challenge of achieving global forest management responsibility through certification is that participation is most feasible in jurisdictions that have well-established forest institutions yet is most required in regions that lack this capacity. Consumer countries are therefore incorporating a "phased approach" in the design of their timber procurement policies in order to lower the barriers to certification adoption in developing regions with weaker regulatory institutions.[33]

An additional source of potential trade distortion is the lack of harmonization of timber procurement policies, particularly among European Union member countries. In an attempt to accommodate the range of local stakeholder interests, governments in different countries have interpreted the various PEFC and FSC certification programs differently, resulting in a form of technical trade barrier and thus a potential distortion of forest product trade flow. In response, governments have been advocating mutual recognition of PEFC and FSC standards and thus greater harmonization of procurement standards.

Disincentive to Sustainable Forest Management

On the one hand, certification can aid in the implementation of forest laws by requiring legal compliance. On the other hand, by raising the cost of market access, there is the potential for it to create perverse market effects – actually encouraging illegal logging and discouraging improvements in sustainable forest management, particularly in developing regions and among small forestland owners. Creating forest management performance requirements that are too high and too rigid simply excludes the marginal forest operators rather than encouraging greater legality and SFM improvements.

In response, as mentioned above, governments are pursuing incremental approaches that phase in legal and sustainable certified timber requirements. Governments are also supporting the development of alternative certification options for small forest operators (for example, FSC and PEFC efforts

Figure 3.9

The spectrum of government's role in forest certification

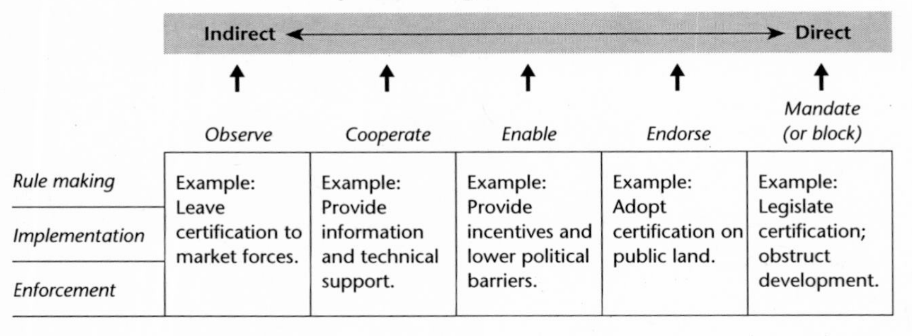

to establish group certification programs to reduce marginal costs). As well, governments have supported the development of national and regional forest certification standards in order to ensure that local sustainability challenges are accommodated.

Indirect and Direct Government Certification Approaches

Depending on the context, governments can take various approaches to certification. As outlined in Chapter 2, government response to a CSR initiative such as forest certification can vary along a spectrum from not taking any action at all to cooperating, enabling, endorsing, and even mandating or blocking forest certification (Figure 3.9). Government positioning can also differ depending on the stage in the policy cycle.

The danger if a government completely ignores certification (leaving it to market forces) is that private forest rules will unintentionally discriminate against certain landowners, create trade disadvantages for certain forest regions, and take forest management in a direction contrary to state objectives – with adverse consequences for the sustainability of the forest and/or local community. Also, if governments of states with weak forest laws and low administrative functioning ignore certification, they may be placing forest policy agenda-setting and rule-making authority completely in the hands of unaccountable organizations, substituting private for public authority and subverting state sovereignty. In order to avoid these worst-case scenarios, a government's approach will therefore necessarily involve engagement in certification with varying indirect to direct responses at the rule-making, implementation, and enforcement stages of the certification process.

Government Positions on Certification

Beyond the need to provide a legal framework and act as a watchdog to ensure policy alignment and prevent market distortions, government practices and attitudes towards forest certification differ within and between countries and regions. For example, government officials presented a range of positions at the 2005 UNECE certification policy forum (Table 3.7). Developed countries (such as the US, Canada, Sweden, Austria, and Germany) defined their role in terms of non-interference, confining government's role to ensuring an appropriate legal framework, providing necessary information for guiding certification alignment with government forest policy, and establishing procurement policies. On the other hand, developing and transitioning countries (such as Brazil, Ghana, Malaysia, and Russia) emphasized the significance of public capacity and a government role in developing and supporting the implementation of certification programs.

Ultimately, government positioning and response to certification is contextual. It will vary by region and according to the respective forest governance conditions. For example, the state may play a greater or lesser role in certification depending on whether it is a weak or strong state, or a producer or consumer country, and depending on the characteristics of the domestic forest policy regime (for example, reliance on prescriptive versus voluntary regulatory approaches, and the extent of public versus private land tenure). New governance arrangements are inherently complex and shaped by existing patterns in different countries. The case studies in the next three chapters focus on the three critical jurisdictions where certification co-regulation is occurring: the highly regulated top global forest product producing nations of Canada, the United States, and Sweden.

Summary

Forest certification emerged as a form of transnational private hard law in response to failed state-led efforts to establish a binding international forest convention. By reviewing and evaluating the development and evolution of forest certification, this chapter has made four key initial points. First, approximately 90 percent of certification uptake has been in developed regions with well-established and enforced forest law; it is therefore in these industrialized regions where the most heightened and evolved interaction between public and private authority in forest policy is occurring. Second, although market actors are a driver of certification adoption, governments have also played a role in encouraging such uptake, and the government role is becoming increasingly important to the ongoing legitimacy of certification programs and maintenance of certified forest. Third, the global private forest regime includes *both* FSC and PEFC programs, whose ongoing competition for legitimacy drives certification evolution. And finally, forest certification constitutes a complementary governance mechanism within

Table 3.7

Summary of government positions on forest certification, 2005

Country	Representative	Government role in certification
Sweden	National Board of Forestry	"Certification is a voluntary agreement between buyers and producers so government has no role in certification but there is interaction between the National Forest Process and certification."
USA	US Forest Service, US Department of Agriculture	"The government does not intervene in certification. It does not act as a standard setting or accreditation body nor does it favor any one certification scheme."
Canada	Canadian Forest Service, Natural Resources Canada	"The federal government views certification as a business decision. The provinces take different approaches to certification. Most leave the matter to individual companies. The federal government is drafting a timber procurement policy which will likely include reference to certified products but not to a particular system."
Norway	Department of Forest and Natural Resources Policy, Ministry of Agriculture and Food	"Government representatives participated in the development of the Living Forest national standard with the role of promoting the C&I [criteria and indicators] of the MCPFE as a reference to ensure that forest certification in Norway would be in accordance with the SFM policy developed at the European level, as well as nationally."
Germany	Federal Centre for Forestry Products	"Government role is to set the legal framework and establish timber procurement policies."
France	International Timber Affairs, Ministry of Agriculture and Fisheries	"Forest certification is a voluntary private initiative complementary to public policies for SFM. The government has certified 100% of the state forests in France and government role in certification is to draw up procurement policies."
Austria	Forest Policy and Information Division, Federal Ministry for Agriculture, Forestry, Environment and Water Management	"The government does not interfere in certification activities. Its role is confined to setting up the appropriate legal framework and providing information necessary for guiding management and certification. Certification is market-oriented and best carried out by the private sector and business community. The government should be attentive to preventing

►

◀ *Table 3.7*

Country	Representative	Government role in certification
		market distortions and build capacity for certification as long as it does not lead to market distortions. As well, certification should be the subject for State-owned forests."
Russia	North Forest Research Institute, Federal Forest Agency	"There is a significant government role to support certification as a mechanism to ensure SFM. The government is involved in the development of a national standard."
Czech Republic	Czech Republic Forest Management Institute	"Forest certification is a private business and the government has no role in it. The government does not support a particular scheme and it is interested in promoting mutual recognition between schemes."
Malaysia	Malaysian Timber Certification Council	"The government has played a significant role in certification. The MTCC was set up and funded by the state."
Brazil	Permanent Mission of Brazil	"The government initiated the certification process in Brazil; helps communities build capability to implement the schemes; and ensures civil society participation ... Governments have role in assuring certification can be an important tool to this end."
Latvia	Ministry of Agriculture	"The government has accepted the FSC scheme to certify its state forests based on UK customer preference. The FSC invited the government to participate in the Latvian FSC meetings. The government views certification as a tool that only verifies that a forest is sustainable. Certification does not bring sustainability."
Ghana	Ghana Forestry Commission	"Governments could cover the full costs of certification for the first five years ... As well, government could address common problems within the various SFM regional processes through engagement with ENGOs in the process of development of harmonized national standards. And they could set up reliable national systems to control possible dilution of certified products with non-certified fiber."

Source: Koleva 2006.

the global forest regime, and a potentially supplementary co-regulatory mechanism within the domestic forest policy mix.

While certification is a unique private governance mechanism (in terms of gaining legitimacy through market supply chains and having an independent private enforcement mechanism), it is neither purely market-driven nor a purely non-state mechanism. Certification standards overlap with public authority, rely on regulatory frameworks, and embed international forest principles and legal compliance. As well, governments are directly engaging in certification. Rather than replacing an existing mechanism, certification can provide an additional decision-making and forest policy delivery instrument for the government's policy toolkit.

Governments face the challenge of determining their optimal response to certification so as to minimize any adverse economic or social effects and maximize potential benefits to the forest, forest owners, forest economy, and local forest communities. Ultimately, certification uptake and outcomes will be determined by the strength of the baseline regulatory framework and how government positions itself.

Overall, although the classification of certification as a non-state market-driven mechanism captures its private governance distinction, it is not an adequate label. Not only do public and private authority overlap within forest certification governance systems but governments are also driving and leveraging these CSR standards. It is important to appreciate and better understand the public/private dynamic of the role of the state in certification systems since optimal co-regulatory approaches will facilitate private rule-making innovation and potentially enhance adaptive governance capacity to achieve sustainable forest management solutions. The next three chapters present the evidence: forest certification co-regulation as it has emerged and evolved over the past fifteen years in Canada, the United States, and Sweden.

4
Canada: Government Authority in Forest Certification

Canada is a global leader in forest certification. There is more independently certified forest area in Canada than in any other nation. Canada is also an international leader in forest conservation policy. So, if forest regulations are already strong and well established in Canada, what forest management role is certification serving? Are private certification systems supplementing or subverting state policy-making authority?

This chapter evaluates the response of Canadian provincial governments to forest certification (1993-2008) and, through the evidence presented, argues that certification did not result in a retreat of government but rather the direct engagement of government authority. Specifically, across similar subnational forest policy regimes (high public land ownership, industrial forestry, and strong forest regulation), provincial governments co-regulated certification by encouraging and participating in standards development, enabling implementation, and mandating forest certification on public land. As well, although the reasons *why* governments engaged in certification were similar as a result of similar forest regime conditions, the differences in *how* governments responded to certification were influenced by three key factors that played out differently in each region: (1) industry expectations of government role, (2) ENGO advocacy pressure, and (3) certification/policy alignment according to the stage of the policy cycle.

Drawing on interviews I conducted in 2004-05 with more than forty-five Canadian forest certification experts and practitioners (see Appendix A), the chapter addresses three central questions:

- Why did provincial governments participate in certification?
- How did governments become engaged at the standards development, implementation, and enforcement stages of certification?
- What factors influenced each government's unique certification response?

The chapter begins with a brief overview of forestry and forest certification development and adoption in Canada and a comparative summary of the key factors that define the forestry sectors of the four provinces included in the study. Each province is then assessed in terms of the history of certification uptake; forest company certification adoption; and the government's response to certification. The assessment then turns to a comparative analysis of the cases. Drawing on the analytical tools presented in Chapters 2 and 3, I first map and analyze the spectrum of provincial government engagement in forest certification, and then identify and evaluate the conditions and factors that influenced provincial government certification response and contributed to the variation in co-regulatory approach between jurisdictions.

Forestry and Certification in Canada

Canada has vast, diverse forest regions (only Russia and Brazil have greater total forest area) that account for one-fifth of the world's temperate rainforest, over one-third of the world's boreal forest, and 25 percent of the planet's remaining primary forest. Beyond the country's unique natural endowment, five key aspects characterize forestry in Canada: (1) the forests are 93 percent publicly owned; (2) provincial governments have the exclusive power and constitutional authority to legislate forest management on Crown (public) forestland; (3) provincial governments delegate responsibility for management of public forestland to the private sector by means of long-term licensing agreements called *Crown forest tenures;* (4) 90 percent of the timber harvested in Canada occurs within old-growth forest; and (5) Canada is the world's largest exporter of forest products, with over 80 percent going to its southern neighbour, the United States. All of these factors have contributed to Canada's enthusiastic participation in forest certification.

Certification Development and Adoption in Canada: National Level

Canada has approximately 138 million hectares of certified forest, over six times the amount of any other nation with the exception of the United States. Sustainable forest management (SFM) certification adoption in Canada has been rapid, largely occurring between 2002 and 2005 (Figure 4.1). In January 2002, the member companies of the Forest Products Association of Canada (FPAC) committed to certifying all of their forest operations by the end of 2006. This goal was achieved and FPAC members now account for approximately 80 percent of all certifications in Canada. Several companies certified initially to the ISO 14001 environmental management system standard prior to seeking SFM certification.

The majority of certification adoption in Canada has been to the Canadian CAN/CSA-Z809 sustainable forest management (SFM) standard, established in 1996.[1] The Canadian forest industry initiated the Canadian Standards

Figure 4.1

Forest certification uptake in Canada, 1999-2007

Canada has 309.8 million hectares of forestland. Of this, 294.7 million hectares are not reserved and are therefore potentially available for commercial forest activities. Of the 294.7 million hectares, 144.6 million are considered accessible, and are therefore most likely subject to forest management activities such as certification. See Natural Resources Canada 2007.

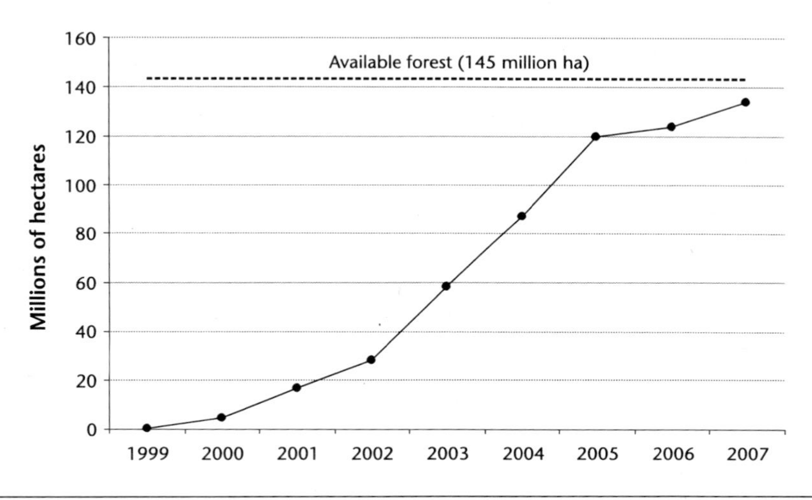

Source: FPAC 2007.

Association (CSA) SFM standard-setting process in October 1993 in competitive response to the Forest Stewardship Council (FSC).[2] The industry feared that the FSC would gain a monopoly of the market and impose overly prescriptive environmental requirements on forest companies, particularly regarding clearcut logging practices and harvesting in old-growth forests. The federal government was also concerned about the economic and international trade implications of the FSC with respect to the standard's encouragement of potential market discrimination towards Canadian forest products. The Canadian industry reacted quickly. In October 1993, while FSC International was holding its inaugural meeting, the Canadian Pulp and Paper Association (now called FPAC[3]) offered the CSA a $1 million contract to develop SFM standards for the industry. The CSA agreed and established a technical committee that had its first meeting in July 1994. Environmental organizations agreed to participate in the multi-stakeholder CSA-led process as members of the Technical Committee. The leading groups (WWF Canada and the Sierra Club of Canada) soon withdrew their support, however, when it became apparent that the standard was being developed in competition with the FSC and would be more focused on process-based management system elements than prescriptive SFM requirements. The CSA standard was released in October 1996.

FSC Canada was established in January 1996. The first Canadian regional FSC standard (for the Maritime provinces) was completed in 1999 and revised in 2008. Other FSC regional standards are the National Boreal standard (August 2004); the FSC British Columbia regional standard (October 2005); and the Great Lakes/St. Lawrence standard (under development). The American Sustainable Forestry Initiative (SFI) standard became a certification option in Canada in 2001 and accounts for approximately 25 percent of the certified forest area across Canada. The CAN/CSA-Z809 standard accounts for 60 percent and the FSC for 15 percent of Canada's certified forest area.

Role of the Federal Government in Forest Certification

The provinces have constitutional authority for managing Canada's public forestland. The federal government's role in forest management is limited primarily to national-level concerns such as trade and commerce, international relations, science and technology, federal lands and parks, and Aboriginal affairs. The federal government took an interest in certification largely because of its potential trade implications, and also to ensure that certification requirements aligned with Canada's national SFM criteria and indicators as incorporated in the country's National Forest Strategy, which was adopted in 1992.[4] It took a direct role in the initiation and development of the CAN/CSA-Z809 standard, and provided funding through the Standards Council of Canada. Its actions followed from the commitment in the National Forest Strategy to develop a national certification system within five years.[5]

In 1998, the Canadian Council of Forest Ministers (CCFM) established a forest certification working group to ensure that certification systems were applicable to the Canadian context, fit within fair international standards, and were not used in foreign markets as discriminatory trade barriers. This was followed in March 1999 by the release of the Canadian Forest Service report *Forest Certification: A Canadian Governmental Perspective,* which summarized the provincial and federal governments' CCFM certification discussions. The document affirmed that governments in Canada supported forest certification as a tool for demonstrating Canada's SFM record, and that Canadian governments were neutral towards the various SFM standards rather than supporting one system over another.

In June 2000, the federal parliamentary Standing Committee on Natural Resources and Government Operations tabled its final report, *Forest Management Practices in Canada as an International Trade Issue* (Volpe 2000). For over two years, the committee had been investigating the linkage between Canadian forest management practices and Canadian forest product exports. Five of the ten committee recommendations pertained to the role of government in forest certification. In response to concerns over fair competition, the committee found that there should be several recognized certification systems available in Canada and that each should respect the principles of openness, transparency, accountability, and equity. The committee also recommended that in cooperation with the provinces and territories, the federal government should encourage the training of SFM certifiers, ensure the maintenance of the policy-making and regulatory functions of governments and international institutions, and promote the international mutual recognition of certification systems.

In summary, the federal government played a direct enabling role in certification by supporting the development of the national CAN/CSA-Z809 certification standard. It also played a key influencing role as the Canadian Forest Service, the CCFM committee, and the recommendations of the 2001 parliamentary Standing Committee all guided the provincial governments in their certification responses.

Provincial Forestry Administration and Certification Uptake

British Columbia, Ontario, and Quebec are the largest forested and timber-producing regions in Canada. Although much smaller, New Brunswick is also a significant forestry region, as its economy is the most dependent upon the forest sector as a percentage of gross domestic product (GDP). All provinces have comprehensive forest law frameworks that enable and support sustainable forest management. All provinces also delegate most Crown forest management responsibilities to private operators through forest license arrangements. Table 4.1 provides a provincial comparative summary.

Most (80 percent) of Canada's certified forest area is located in British Columbia, Ontario, Quebec, and New Brunswick, in decreasing order. In 2004, New Brunswick had certified the greatest percentage of its total productive forest (78 percent) and Quebec the least (8 percent). Certification has occurred rapidly in each province over the past several years. Companies in BC were the first on board with SFM certification. For example, MacMillan Bloedel and Weldwood certified to the CAN/CSA-Z809 standard in 1999 and 2000, respectively. Irving Forest Products in New Brunswick was an early adopter of the FSC, certifying to the FSC International principles in 1997. Ontario has been the leader in terms of total FSC-certified forest, with Tembec and Domtar's FSC certifications accounting for 35 percent of Ontario's certified forest area.[6] Certification in Quebec has lagged behind that in the other provinces, with most forests being certified in 2005-06.

The focus here is on BC, Ontario, Quebec, and New Brunswick not only because these provinces represent coast-to-coast coverage of the major forested and forestry producing regions in Canada but also because they provide a representative spectrum of varying government and industry engagement in forest certification. Although the provinces are relatively similar in terms of their industrialized forest economies and prescriptive forest policy regimes,

Table 4.1

Comparison of national and provincial forestry characteristics in Canada

Characteristic	Quebec	New Brunswick	Ontario	BC	Canada
Population (2004) (millions)	7.5	0.75	12.3	4.2	31.8
Total land area (millions of hectares)	136.5	7.1	91.8	92.5	909.4
Total productive forest area (millions of hectares)	43.6	6.1	22.0	60.6	309.5
Provincial forest ownership	89%	49%	91%	96%	93%
Harvest (2002) (million cubic metres per year)	39.6	10.1	26.3	73.6	189.2
Value of forest exports (2003) ($ billions)	10.7	2.3	8.5	12.6	39.6
Contribution to provincial GDP (2004)	4%	8.6%	1.4%	7.2%	–
Certified forest (December 2003) (millions of hectares)	6.6	4.7	13.2	29.3	57.7
Certified forest (January 2008) (millions of hectares)	17.7	4.3	26.3	51.5	137.9
Provincial forest legislation	Forest Act (1986, revised 2001)	Crown Lands and Forest Act (1982)	Crown Forest Sustainability Act (1994)	Forest and Range Practices Act (2004)	–

Sources: CSFCC 2008; Natural Resources Canada 2004a.

they have positioned themselves differently with respect to their roles in certification. For example, both New Brunswick and Ontario have mandated certification (on public land), whereas BC and Quebec have taken different co-regulatory approaches (Figure 4.2).

Figure 4.2

Certification approaches taken by provincial governments

British Columbia

From the early stages of certification development through the 1990s, the BC provincial government took a relatively passive role with respect to certification implementation and enforcement, but participated directly in the development and promotion of the CSA national SFM standard. Over the past decade, as certification programs have gained in uptake and legitimacy, the government's role in certification has been shifting. The BC Ministry of Forests and Range has become more directly engaged in the FSC regional standard-setting process, has been encouraging the alignment of certification with the new results-based forest legislation, and has also supported certification of the provincial timber sales organization (BC Timber Sales). British Columbia was an early proponent and adopter of certification, and plays a leading role in certification in Canada and the world. In 2008, BC was the top region in North America in terms of total certified forest area.

The Provincial Context

Just over 40 percent of Canada's timber volume is located in British Columbia and the province produces over one-half of the nation's lumber exports. Historically, large industrial forest companies have accounted for approximately 80 percent of the provincial timber harvest and operate primarily on public land. Virtually all (95 percent) of BC's forests are publicly owned and under strict forest law. The Ministry of Forests and Range is the main agency responsible for protecting and managing the 60 million hectares of provincial forestland as well as providing the basis for a globally competitive forest industry with high environmental standards and maximizing net revenues to the Crown.[7] The province sets the allowable annual harvest level for Crown land and delegates forest management responsibilities to forest licensees largely through volume-based tenure arrangements.

BC is also home to a stunningly unique global ecological endowment. This western Canadian province has a greater diversity of temperate and boreal forest types than any other jurisdiction in Canada or the US, including 20 percent of the world's remaining temperate rainforest. With iconic thousand-year-old trees towering over 80 metres in height and measuring up to 19 metres in circumference, these undisturbed ancient coastal forests have been the focus of global campaigns to stop their destruction from clearcutting industrial logging practices. International preservation battles have played out in BC since the early 1980s, particularly in the Pacific coastal regions of South Moresby Island (Gwaii Haanas) (1985), Carmanah Valley (1989), Clayoquot Sound (1993), and the Great Bear Rainforest (1995-2006). In the 1980s and 1990s, devastating media images of slashed, scarred, and eroding BC forest landscapes shocked the global community and led to the perception of regulatory failure in BC.

Timber-based industries are the foundation of the provincial economy, and the environmental campaigns against the BC forest sector were therefore of concern to every citizen in the province. Market access, jobs, and prosperity were threatened. The government responded in the early 1990s by revamping the province's forest legislation, establishing a Protected Areas Strategy (PAS), conducting a timber supply review, and initiating province-wide land-use planning public consultation processes (Commission on Resources and Environment [CORE] and Land and Resource Management Plan [LRMP]).[8] The government also introduced a new Forest Practices Code of British Columbia Act (1994) and very detailed accompanying forest management regulations (released in 1995). The new forest policy regime constituted a much more stringent prescriptive approach to forest practices, planning, and enforcement than the government's previous policy approach, which had relied for the most part on the inclusion of contractual obligations and voluntary guidelines in forest operational plans and permits.

Fundamentally, certification was successfully adopted in BC because both the forest industry and the government faced the same challenge. Both needed to rebuild domestic and international trust in the province's implementation and enforcement of sound sustainable forest management practices. Independent third-party certification was viewed as an opportunity to reinforce recently revised forest legislation and win back market and social confidence in the province's forest sector.

Forest Certification Uptake in British Columbia

Certification in British Columbia occurred rapidly, increasing from 210,000 hectares in May 1999 to just over 50 million hectares at the end of 2007. Industrial companies hold over 90 percent of the SFM certifications issued in the province (Table 4.2). The largest certification holder in the province is Canfor, followed by West Fraser Timber and the BC Timber Sales program (BCTS). As shown in Table 4.2, the majority of certified forest has been to the CAN/CSA-Z809 standard. As of 2008, only one *major* operator in BC (Tembec) has obtained FSC certification, accounting for 90 percent of the provincial FSC total. The other significant FSC holder is the First Nations–owned Iisaak Forest Resources, which FSC-certified 87,393 hectares of coastal old-growth forest in Clayoquot Sound in 2001.[9]

Forest Company Certification Response in BC

Overall, BC forest companies were leaders in initiating and supporting the CSA standard but laggards with respect to FSC acceptance and adoption. Given the history of conflict with environmental groups, BC forest companies were initially afraid of the economic consequences of the ENGO-led FSC and immediately positioned themselves as proponents of the competing national certification scheme, the CAN/CSA-Z809 SFM standard.

BC companies were the first in Canada to achieve CSA forest certification.[10] By the end of 2003, every major BC forest company either had certified or was in the process of obtaining forest certification. The biggest obstacle to CSA certification in the province was figuring out how to certify a defined forest area under a volume-based tenure arrangement.[11] Companies were uncertain how to address shared licensee responsibilities, and were also unclear as to how a licensee could apply for certification without having the provincial government as a co-applicant (given that the government owned the land and was responsible for setting the harvest level). In the absence of clear direction, several companies focused initially on International Organization for Standardization (ISO) certification of their facilities (rather than the forest) and SFM certification of their smaller area-based tree farm licences (TFLs).

Table 4.2

Major forest certification holders in British Columbia, January 2008

Standard	Major licensees	Certified area (ha)	% of provincial certification	Date of initial certification(s)	British Columbia total certified (ha)
FSC	Tembec	564,776	1.0	Nov 2004 – Sep 2006	
					FSC total: 577,295
CSA	AbitibiBowater	2,132,736	4.0	Dec 2004	
	Ainsworth	887,194	1.7	Dec 2004	
	BC Timber Sales	4,014,110	7.8	Apr 2005 – Aug 2007	
	Canfor	15,443,833	30.0	Jul 2000	
	Tolko	3,406,927	6.6	Mar 2003	
	Fort St. John	2,550,000	5.0	Oct 2003	
	Western Forest Products	904,528	1.8	May 1999	
	Weyerhaeuser	954,000	1.9	Mar 2001	
					CSA total: 33.1 million
SFI	BC Timber Sales	4,585,261	8.9	Feb – Dec 2007	
	Interfor	2,229,073	4.3	Jan 2001 – May 2004	
	Louisiana- Pacific	3,016,750	5.8	Sep 2001 – Jul 2005	
	Pope&Talbot	1,173,588	2.3	Nov 2002 – Aug 2005	
	Timberwest	482,293	1.0	Dec 2000 – Dec 2007	
	West Fraser	5,100,000	10.0	Dec 2001 – Nov 2005	
					SFI total: 17.8 million
					British Columbia total: 51.5 million

Source: CSFCC 2008.

Although there were several issues with the FSC, the key operational reasons that the standard achieved very little traction in BC were FSC principle number 3, which recognized "the legal and customary rights of indigenous peoples to own, use and manage their lands, territories, and resources," and principle number 9 regarding the maintenance of primary forest. Given that the majority of First Nations land claims in BC were unresolved and the majority of timber harvested in BC consisted of old-growth forest, the industry anticipated that the FSC would severely restrict forest access and lead to dire economic consequences. For example, Canfor conducted an FSC pilot on its Dawson Creek tree farm licence and determined that FSC certification would reduce its harvest by 35 percent.

Despite the risk of lost market access resulting from ENGO boycotts and customer pressures,[12] BC forest companies did not accept or pursue FSC certification. Instead, they directed their attention to developing an ecosystem-based management strategy for coastal BC under the Joint Solutions Project (JSP).[13] The delay in establishing a regional FSC standard reflected the defensive industry position. Although a regional FSC office (FSC-BC) was established in 1996, it took almost ten years before a regional standard was approved in 2005.

The BC Government's Certification Response

Position on Certification

When certification emerged in 1993, the BC government was preoccupied with the development and implementation of new forest legislation, regulations, and comprehensive forest management programs, and was immediately wary of the FSC private forest governance scheme. The government believed that a negotiated consensus on sustainable forest management values and objectives had already been achieved with the citizens of the province through land-use planning, and did not want the issues reopened, particularly under ENGO direction. As Don Wright, the Assistant Deputy Minister of Forests at the time explained, "The government had already invested a lot of dollars in a green agenda and had struck a fair balance with the Forest Practices Code and the Protected Areas Strategy. FSC was the ENGO agenda not the provincial agenda."[14]

Absorbed in the challenges of implementing provincial programs and new regulations, the government viewed the FSC's private governance rules as a competitive threat to its forest policy agenda. It also acknowledged, however, that third-party independent certification was an opportunity to rebuild confidence in the BC forest sector, which had been shaken by the international ENGO boycotts and protests. The government's position was therefore to support certification as long as it was appropriate to the BC context and aligned with the province's forest legislation.

Initially, the government hoped that it could counter the negative global campaigns and European boycotts against BC forest products and set the record straight with offshore customers by simply communicating BC's proactive sustainable forest management laws and practices better. The Intergovernmental Affairs Group (working out of the premier's office) initiated a European delegation so that the premier could meet with overseas buyers, governments, and media to promote BC forest practices and the province's new forest policy initiatives (BC Ministry of Forests 1998). The European meetings proved largely ineffective, however. Officials came home with the realization that regulatory reforms went only so far in addressing the concerns of the international ENGO campaigns and restoring international market trust. A more integrated strategy that involved environmental and First Nations organizations as well as the federal government was required. As Premier Ujjal Dosanjh announced in October 2000, "market challenges for BC forest products require an integrated, collaborative response ... we must work together ... to make sure we continue to supply the world with the highest quality wood products available anywhere."

Although the government had major concerns about the FSC (estimating that the annual cut would be reduced by over 40 percent), it was cautious about taking sides between the CSA and FSC programs. The government formally adopted a position of "passive neutrality." As Don Wright explains, "the government was torn between two constituencies ... on the one side there was industry and the CSA standard, and on the other side the ENGOs and FSC ... the government wanted good relations with the ENGOs but also needed industry prosperity ... the government felt that if they supported CSA they would lose ENGO support and if they supported FSC they would discourage industry ... therefore, the government decided to support the *principle* of certification."[15]

From 1995 to 2000, certification gained increasing government attention. For example, the Ministry of Forests' 1995/96 annual report mentioned certification in just one sentence, about the province's work on developing a long-term forest vision: "The ministry will be participating in domestic and global efforts to develop an internationally acceptable system of certifying sustainable forest practices." A few years later, however, the ministry's 1999/2000 annual report dedicated a full page to outlining the government's Certification of Forest Products Initiative, an integrated collaborative arrangement with the Ministry of Employment and Investment and the Ministry of Environment, Lands and Parks to "ensure that provincial government interests are properly factored into certification strategies."

From 2001 to 2005, the forest policy regime in BC shifted from a prescriptive to a results-based regulatory approach.[16] Over this period, the government continued to demonstrate increasing engagement in and acceptance of certification, not just through BCTS certification but also in exploring the

Figure 4.3

Positive and negative policy values of certification

How can certification add value?	*How can certification subtract value?*
• Serve as international communication tool	• Privatize "policy" issues such as land use, protected areas, zoning
• Operationalize "multiple-value" forestry at each enterprise level	• Add costs beyond market benefits
• Create other policy options besides legislation	• Create unpredictable forestry requirements
• Support environmental standards through market benefits	• Create inequities and market access problems due to varying standards
• Provide positive market incentives towards best practices	• Diminish public involvement or say in forest policy issues

Source: den Hertog 2000, 5-6.

alignment of certification with forest policy and programs. In 2002, the government formally stated its position on certification: "As a market instrument, certification operates outside of the regulatory framework established by governments. However, the BC government has a specific interest in certification because it has the inherent potential to affect access to markets, reinforce sustainability requirements for forest management and support or contradict domestic and international legislative and policy goals" (BC Ministry of Forests 2002b).

Overall, the BC government viewed certification as an additional policy tool that, depending on the government's role, could either add or subtract value with respect to the achievement of the province's SFM goals (Figure 4.3). The government therefore pursued a co-regulatory approach with the aim of maximizing the positive and minimizing the negative potential policy impacts of certification.

Role in Certification Development

When forest companies in BC and the Canadian Pulp and Paper Association initiated the development of the CSA national standard in 1993, the BC government lent its support and became a member of and active participant on the CSA Technical Committee. The province had already been working with the federal government and the other provinces through the Canadian Council of Forest Ministers to develop Canada's national set of criteria and indicators for sustainable forest management. These national criteria formed the basis of the CSA standard. Although not formally acknowledged or communicated by the participating governments, the CSA standard essentially took the CCFM criteria and indicators from the development to the implementation stage – giving them "legs to walk on."[17] In addition, the BC

Ministry of Forests contributed to Canada's national advisory committee on ISO environmental and labelling certification standards (BC Ministry of Forests 2000b, 27).

Although cautious about the FSC, the provincial government also participated as a non-voting member in the FSC-BC standard-setting process, arguing that it was better to be involved at the table at the start than after the fact.[18] The government also participated in federal delegations to Europe, the US, China, and Japan to market Canadian forest products and publicize the fact that the CSA SFM program conformed to international SFM criteria and met ENGO concerns.

Unlike many forest producing regions that focused on the supply side in terms of facilitating an increase in the volume of available certified fibre, the BC government put greater emphasis on the demand side – actively promoting and marketing BC forest practices and BC certified wood to export customers.[19] In October 2000, Minister of Forests Jim Doyle explained that "our work with industry and certification groups helps us respond to buyers who are increasingly seeking certified wood products ... but we also need to make sure our customers know BC's record of conservation improvements to our forest practices over the past decade ... we will communicate BC's record of producing quality products using quality forestry methods and BC's commitment to doing an even better job" (BC Ministry of Forests 2000c). Careful to protect the government's policy authority, the message consistently stressed the importance of communicating the province's regulatory strength as well as the alignment of third-party independent certification with forest policy in BC.

Role in Certification Implementation

The BC government was directly engaged in the development of the FSC-BC and CSA standards but took a less indirect, passive role in the implementation of the standards, encouraging licensees to certify but neither endorsing nor creating incentives for the adoption of a particular standard. BC companies were generally frustrated with the government's lack of leadership in helping to overcome some of the initial certification hurdles, particularly in choosing between the various standards and regarding the certification of volume-based tenures. Some companies described the government as a "fence-sitter."[20]

In order to track certification and facilitate adoption, the government hired a certification implementation coordinator, established a dedicated unit to address certification issues and monitor certification developments worldwide, and designated a contact in each of the ministry's forty-six regional and district offices to provide information and assistance to licensees applying for certification. For the most part, the government delegated its certification implementation role to the regional and district offices. As explained by several BC companies, the head office in Victoria observed and the districts led. District Ministry of Forests, Ministry of Environment, and federal Department of Fisheries and Oceans (DFO) staff participated in and provided technical guidance to local CSA certification public advisory groups. The level of government support varied not only according to district but also depending on the individuals involved. For example, local district office employees were particularly enthusiastic in providing the Dawson Creek and Fort St. John certification projects with a high level of support.

A key government concern with respect to certification implementation was to ensure the applicability of certification programs to BC and their alignment with provincial, national, and international forest policy and agreements and with the province's forest policy agenda. Specific government initiatives involved working cooperatively with forest licensees to pilot the CSA and FSC certification standards in several regions.[21] The pilots also tested certification with the province's small business program in these regions. A few months later, in June 2000, the government appointed a thirteen-member multi-stakeholder advisory council on certification. As newly appointed Minister of Forests Doyle announced, "I have asked representatives from the forest sector, First Nations, environmental groups, labour, and local communities to provide advice on implementing certification in BC quickly and efficiently ... we need to work together and identify how certification can work to support our economy and protect our environment" (BC Ministry of Forests 2000a). In the spring of 2000, the provincial government was announcing certification implementation as a key policy priority.

In the same announcement, the government commissioned a study to assess the issues and options related to its role in forest certification in British Columbia. Based on input from stakeholders across the province, the consultant's report concluded that "government should work co-operatively with licensees and other interests in implementing forest certification but should remain neutral on the merits of alternative certification systems" (BC Ministry of Forests 2001). The report also recommended fifteen actions for the government over the next two to three years. These ranged from providing information, training, and technical advice to facilitate certification uptake to encouraging greater integration and alignment of certification and forest policy (Brown and Greer 2001).

The government's interest in enabling certification was reinforced with the agenda of the new Liberal government (elected in 2001) to streamline provincial regulation, and with the enactment of the results-based Forest and Range Practices Act (FRPA) in 2002. For example, in the spring of 2004, Dr. John Innes at the University of British Columbia was contracted to study the extent of the relationship between evaluations and monitoring conducted under the FRPA versus the monitoring and assessment of SFM practices under forest certification. The government was interested in the co-regulatory opportunity to integrate the private governance system into the government's policy mix, but was not clear on the compatibility of the two systems. While the report outlined a range of difficulties, it also stressed the significant potential for collaboration and interplay between the public and private systems – "that with further work and analysis the FRPA evaluation process could eventually incorporate certification measures, reducing the province's overall forest monitoring evaluation costs." [22]

Role in Certification Enforcement

The mandating of forest certification was not seriously considered in BC.[23] The government approached forest certification as a voluntary, market-based private regulatory system separate from the traditional regulatory framework. Although it did not support enforcement of certification on Crown land, it did adopt it for its own Small Business Forest Enterprise Program (SBFEP), and eventually encouraged it for all of BC Timber Sales.[24] In 2002, the government implemented a change in legislation that would enable it to enforce certification among small forest operators. In announcing the legislative amendment, Minister of Forests Michael de Jong stated that "the province is committed to sound forest and environmental management practices that are recognized both locally and internationally ... by responding to the market demand for certified wood, we are working to make our forest industry more globally competitive" (BC Ministry of Forests 2002a). By January 2008, the BCTS program had become the third-largest certification holder in the province (see Table 4.2).

The government also looked for ways to leverage certification to lessen the province's regulatory costs. For example, the Forest Practices Board (FPB) investigated the potential to streamline their compliance audits by using certification audit results.[25] The pilot projects revealed that certification audits did not align well with the province's compliance audits,[26] but as of 2005 the board's position was to look at opportunities on a case-by-case basis rather than completely give up on the possibility of using certification to streamline its audits.[27]

In summary, the BC government's approach consistently emphasized the role of certification in supporting, not supplanting, the province's strong regulatory regime. Initially, the government took a hands-off approach to the FSC because the standard's early requirements did not align well with provincial forest policy, and pursued a direct co-regulatory role in the development of the CSA standard to ensure consistency with provincial policy. Over the past decade, the government's role has shifted from taking an indirect cooperative approach towards certification implementation and enforcement to directly engaging in and enabling certification as an additional policy tool.

New Brunswick

On Canada's east coast, New Brunswick is a much smaller timber-producing province relative to Ontario, Quebec, and British Columbia, but it is one of the country's oldest industrial forestry regions and has been a leader of industrial forest management policy. In 1837, it established forestry regulation to protect state timber revenue, and in 1883 it was a provincial leader in introducing forest conservation policy. In 1937, it was the first province to introduce regulations that delegated public forest management silviculture and planning responsibilities to industrial forest operators. In 1966, New Brunswick was also the first province to create an integrated Natural Resources Department, combining responsibilities for forests, minerals, and energy. In April 2002, New Brunswick again demonstrated its forestry policy leadership by becoming the first jurisdiction in North America to mandate forest certification, requiring it of concession holders on provincial Crown land.

The Provincial Context

The provincial government owns just over half of the forestland in New Brunswick; 29 percent is held by 40,000 small private woodlot owners and 18 percent is in industrial freehold.[28] Under the 1982 Crown Lands and Forest Act (CLFA), the provincial Department of Natural Resources (DNR) is responsible for managing the Crown forestlands. The province delegates public forest management responsibility to industrial forest companies through ten twenty-five-year area-based Crown Timber Licences. As of 2002, four companies accounted for 90 percent of the allocated forest licence area: J.D. Irving (32 percent), UPM-Kymmene Miramichi Inc. (29 percent), Fraser Papers (16 percent), and Bowater (13 percent). St. Anne Nackawic Pulp Co. Ltd. and Weyerhaeuser accounted for the remaining 10 percent. Crown land accounts for about 43 percent of the total fibre supply in the province, and 72 percent of this fibre is softwood.[29] Industrial freehold and private woodlots supply approximately 23 percent and 21 percent, respectively. New Brunswick mills demand more fibre than is supplied locally, so the province relies on imports from Maine, Quebec, and Nova Scotia.[30]

The DNR sets Crown forest management goals and objectives, regularly monitors and assesses licensee activities, reviews and renews the licences every five years, and assigns the allowable annual cut. In return for access to public timber, licensees are required to prepare long-term forest management plans and annual operating plans and to meet all government requirements. Smaller mills are allocated Crown timber volume through sublicences. In 2004, there were about eighty sublicensees operating in the province.

There is very little primary forest remaining in New Brunswick, despite the fact that 85 percent of the province remains forested. New Brunswick has the highest forest cover in Canada, with over 6 million hectares of productive second growth. The province's Acadian forest includes mixed northern hardwoods and coniferous species that were regenerated after intensive logging and agricultural clearing in the nineteenth century.

Compared with other provinces, the New Brunswick economy has the greatest dependency on forestry, which historically has accounted for approximately 9 percent of the provincial GDP. Like all other Canadian forest producing regions, the province's forest industry is export-dependent, shipping over 80 percent of its products to US customers. New Brunswick's largest forest sector is pulp and paper, although the province also produces solid wood and manufactured wood products.

Forest Certification Uptake in New Brunswick

All of the major forest certification holders in New Brunswick have certified to the US SFI standard (Table 4.3).[31] Certification occurred rapidly between 2000 and 2003. All Crown forest licensees were certified by 2003. With two exceptions, sublicensees certified under the scope of the major licensee certifications.[32]

Forest Company Certification Response in New Brunswick

J.D. Irving, based in Saint John, was the first forest company in Canada to achieve FSC certification, in 1999. As explained later in this section, however, the company shortly thereafter withdrew from the FSC program due to disagreement with the Maritimes regional standard. UPM-Kymmene was the early certification leader to the ISO and SFI standards. Time Inc., a major customer, had approached the company about providing SFM-certified fibre and the company responded.

Although New Brunswick forest companies had initially intended to certify to the CSA standard (and many had achieved ISO 14001 certification in preparation), all of them ended up pursuing SFI certification. A key reason for this was to facilitate sublicensee certification and provincial wood procurement.[33] Acting on government advice to try to reach agreement on a uniform certification approach, the industry chose the SFI standard. Reaching consensus was fairly straightforward as the industry had a history of working cooperatively to resolve critical provincial forest issues dating back to the spruce budworm infestation in the 1970s.

Companies chose the SFI over the CSA standard for several reasons: (1) at this time (2001), there was little demand for or recognition of the CSA standard; (2) the vast majority of New Brunswick forest products are sold to US customers, so it made sense to certify to an American standard; (3) the province's forest industry relies on fibre from many private woodlot producers and the SFI placed strong emphasis on wood procurement; and (4) companies felt that CSA certification would take too long to achieve.[34] Every company in New Brunswick operates on Crown land under a Forest Management Agreement that requires public consultation on the management plan every

Table 4.3

Major forest certification holders in New Brunswick, January 2008

Standard	Forest operator	Certified area (ha)	% of provincial certification	Date of initial certification(s)	New Brunswick total certified (ha)
FSC	Eel Ground Community Development Centre Inc.	2,853	<1	Sep 2005	
					FSC total: 3,739
SFI	A.V. Kackawic	296,127	7	Dec 2006	
	AbitibiBowater	426,352	10	Nov 2003	
	AT Ltd. Partnership	844,984	20	July 2000	
	J.D. Irving	1,790,813	41	Dec 2000 – Dec 2003	
	UPM-Kymmene Miramichi	942,919	22	Dec 2002	
					SFI total: 4,301,195
					New Brunswick total: 4,304,934

Source: CSFCC 2008.

five years, whereas the CSA's 1996 standard required an ongoing local public advisory group.[35] As the chief forester of Fraser Papers explained, "we had our advisory committees and our management plans in place ... the only hitch with the CSA process was that public participation demanded even more than we had already done. It would have required us stepping back a couple of years and reworking our management plan" (Forest Certification Watch 2004a).

No forest company in New Brunswick sought FSC certification or participated on the FSC Maritimes regional committee because of the heightened politics around the development of the FSC Maritimes regional standard and J.D. Irving's unsuccessful experience with the organization. The industry did not support the standard itself either, perceiving it to be a forest restoration standard rather than a forest management standard.[36] Two small woodlots did achieve FSC certification in 2003.[37]

J.D. Irving and the FSC

In 1998, J.D. Irving FSC-certified 231,000 hectares of its Allagash Woodlands in Maine and 190,000 hectares of its Black Brook forest operation in New Brunswick.[38] The FSC's Maritimes regional standard had not been approved at this point, so the New Brunswick forest was certified to the FSC International principles while the Maine forest was certified to the US North East regional FSC standard. In June 2000, J.D. Irving abandoned its FSC certifications in both New Brunswick and Maine because of its disapproval of the Maritimes standard. The company felt that the standard had been developed without adequate industry representation, lacked scientific basis as it recommended the virtual elimination of biocides, and would create an uneven playing field between New Brunswick and its competing neighbour, the state of Maine. Although the forests in New Brunswick and Maine were very similar, the FSC regional standards were very different. According to J.D. Irving's chief forester, the company cancelled its FSC certification because, "we felt the standards were unreasonable and they didn't have broad stakeholder support. They weren't consistent with other FSC standards in neighbouring regions or anywhere else in the world" (Forest Certification Watch 2004a).

The New Brunswick Government's Certification Response

Position on Certification

The New Brunswick government's initial approach to certification (1996-2000) was simply to observe and learn. It participated in the federal government's CCFM certification committee and engaged in discussions with other provincial governments to figure out their position on certification. Before taking on any sort of formal role, it wanted to wait and see what the New

Brunswick forest industry was going to do. The DNR initiated dialogue with the industry and learned that companies such as UPM-Kymmene were pursuing certification to the ISO and SFI standards and that industry perceived certification benefits to include market access and continual forest management improvement. The government also closely observed J.D. Irving's difficulties with its FSC certification, and monitored the battles taking place between the various certification systems. In the end, it clearly took a neutral role – not interfering in the J.D. Irving dispute and not taking sides supporting one standard over another. When marketing New Brunswick forest products outside the province, the government did not want to be seen as an FSC opponent and carefully communicated its support for SFM certification in general, not one particular standard. The DNR explained: "The department saw certification as market-driven. We didn't think the government should take a role in promoting one system over another. Market forces shift and industry could end up losing market share and then the government could be liable."[39]

The New Brunswick government was very proud of its Crown land system and history of policy leadership in Canada. After observing and learning about certification and the industry's enthusiasm for provincial certification, the government again took a policy leadership role and became the first jurisdiction in North America to mandate forest certification on its public land. Ultimately, it approached certification as an additional forest management tool that was good for the forest industry and that provided public assurance that the province was managing its forests well. As the DNR noted, "it was better to have a third party independent auditor passing judgment than the Minister saying our forest practices were good. Certification removed the bias."[40]

Role in Certification Development

The government took a cooperative co-regulatory role in standards development. The DNR participated in the CSA Technical Committee and attended the FSC's Maritimes regional standard meetings as an observer. The government also played a supportive role in the Canadian Federation of Woodlot Owners initiative to develop a certification program for private woodlot owners.[41] The DNR's rationale for engaging in the development of the various certification standards was to ensure policy alignment – to prevent conflict between certification and the province's forest policy objectives.

Role in Certification Implementation

Overall, the New Brunswick government also assumed a cooperative role in facilitating certification implementation. During the initial stages of certification adoption in the province, the industry called on the government to help it decide which standard to pursue and to provide guidance on how to address sublicensee certification. The DNR suggested that companies discuss among themselves the opportunities and benefits of perhaps working together under one system. The industry followed this suggestion and adopted a uniform approach, with all companies certifying to the SFI standard.

Specifically, the DNR cooperated in certification implementation by providing technical assistance to the licensees when requested, offering clarification of provincial policy, particularly during certification audits, and participating in the SFI implementation committee (SIC).[42] As well, the government's mandatory certification requirement in 2002 spurred industry's certification implementation efforts.

Although certification of the 40,000 small private woodlots across the province was a challenge,[43] the government did not intervene directly. Instead, it let the forest companies work it out with the many private forest owners. And mentioned earlier, it also supported the efforts of the Canadian Federation of Woodlot Owners to develop and implement a feasible Canadian woodlot certification standard.

Role in Certification Enforcement

Although during the 1990s the New Brunswick government initially took a wait-and-see approach to certification, it announced in April 2002 that ISO 14001 certification would be required for all Crown timber licensees by the end of 2002, and SFI, CSA, or FSC SFM certification by the of 2003. Industry had approached the government about mandating certification on Crown land. Licensees were already certifying, and their forest management plans were already addressing certification requirements.[44] In addition, customers such as Time Inc. were pressing for certified forest products and the industry saw provincial Crown land certification as a means of promoting New Brunswick forest products and meeting increasing market demands. The government saw opportunity in directly co-regulating certification since the costs of implementation were minimal (given that the Crown Lands and Forest Act aligned well with certification requirements), and mandating certification would promote the province's forest management legislation and propel New Brunswick back into a forestry policy leadership position.[45]

In 2004, following up on a Jaakko Pöyry consultant's study on how to increase Crown land wood supply,[46] the government focused on the interaction between certification and forest policy. Specifically, the DNR examined the possibility of redesigning the province's Crown land compliance monitoring program along the lines of a certification audit system. The challenge was to determine ways to reduce overlap between certification auditing and provincial monitoring while "maintaining the custodial role of DNR on Crown land" (NBDNR 2004, 29).

Table 4.4

Options for aligning certification audits and government monitoring in New Brunswick

Option	Description	Implications
Status quo	DNR and licensees do regular field checks. An annual review pinpoints areas of concern. DNR performance evaluation takes place at the end of the planning period.	DNR retains custodial role as effectively and efficiently as possible.
Replace DNR Crown land oversight with certification	One system of verification on Crown land.	Certification will not cover the day-to-day inspections and will leave shortfalls in the inspection process. Certification can enhance but not substitute for the work of experienced DNR field staff.
Co-regulation: certification reduces or reinforces DNR oversight	The licensee would adopt a certification process that incorporates DNR's operational criteria and is implemented by a third-party certifier.	The option would require more time and effort to implement but DNR oversight costs could be decreased, leaving DNR staff with time and flexibility to devote to other activities.

Source: NBDNR 2004, 31.

The province wanted to leverage certification capacity without ceding authority over forest management. The DNR staff review of the Jaakko Pöyry study proposed three options to address the study's recommendation to reduce overlap in licensee and DNR management/supervision of Crown lands (Table 4.4): continuing with status quo DNR audits, substituting DNR audits with certification audits, or establishing a co-regulatory public/private audit process.

To summarize, the New Brunswick government positioned itself in an indirect cooperative role to facilitate certification development and adoption in the province, and a direct role in terms of mandating certification on provincial Crown land. The early difficulties involving J.D. Irving and the FSC Maritimes regional standard caused the government and industry to go into damage-control mode to reassure markets that New Brunswick forest management practices met international standards. Mandating certification was an intentional co-regulatory strategy on the part of the government, in order to promote the province's forest products, win back market trust, and reassert the province's historical forestry policy leadership.

Quebec

Quebec has the greatest total area of productive forest in Canada (14 percent), but the province has lagged in certification adoption. In 2004, Quebec accounted for only 7.6 percent of Canada's total certified forest area. Over the past decade, Quebec forest policy has gone through major reform, and with the introduction of greater public participation and ecosystem-based management considerations, the government also revised its approach to certification. Its role shifted from non-intervention to active facilitation of certification as an additional mechanism in the province's forest policy regime. In December 2007, the Quebec parliament amended its forest legislation to enable the minister of natural resources to mandate certification; as of January 2008, Quebec ranked third in Canada in certified forest area.

The Provincial Context

Forest covers half of the province of Quebec and includes boreal, mixed, and hardwood forest types. Softwood species located largely in the northern part of the province (such as fir, spruce, jack pine, and larch) account for most of the timber volume and annual harvest. The hardwood forests in the southern part of the province are closely tied to the country's national identity, as 90 percent of Canada's maple syrup production (and 70 percent of world production) is from Quebec. Quebec is Canada's leading paper producer and ranks second to British Columbia in logging and wood product manufacturing. The forest sector is a key contributor to the provincial economy, and Quebec has greater total forest industry employment compared with any other province.

The provincial government owns 89 percent of the 49.8 million hectares of total forest area and 84 percent of the 43.6 million hectares of productive forest area. Approximately 130,000 private woodlots account for 5.5 million hectares of the productive forest (13 percent), and large industrial holdings account for 1.1 million hectares (3 percent). Public forests account for 76 percent of the annual 38 million cubic metres of timber harvested.

The Department of Natural Resources and Wildlife (MRNF) is responsible for managing public forests under the provincial Forest Act (established in 1986 and revised in 2001).[47] As in other Canadian provinces, the public forests are divided into management units that are then allocated to forest companies under twenty-five-year Timber Supply Forest Management Agreements (TSFMA). The government assigns the allowable annual cut, monitors licensee activity, and reviews and renews licence agreements every five years. In exchange for access to public fibre, licensees prepare five-year management plans and agree to meet government forest development and protection objectives. In 2004, there were 239 TSFMAs across 114 common areas. Private forests are managed through forest management agreements and municipal laws.

Unlike other provinces, Quebec's Crown land is allocated to forest companies on a shared volume basis. Under the Quebec forest management system, TSFMA holders can "cohabitate" on a forest area. One licensee may be licensed to harvest hardwood while another may be given rights to a certain softwood species. In Quebec, public land licence holders are therefore collectively accountable for the forestry activities on the entire management unit and submit joint five-year and general (twenty-five-year) forest plans. One company is designated to write the plans and present them to the general assembly of licence holders for the management unit. No licensee has veto power and all must agree on the plans. Navigating this shared tenure arrangement has been a significant certification challenge.

Forest Certification Uptake in Quebec

Quebec certification occurred later than in other provinces, with most adoption taking place between 2005 and 2006. Quebec companies have divided their support between the FSC, CSA, and SFI standards (Table 4.5). In terms of forest area, CSA has the largest percentage share (67 percent), followed by FSC (25 percent) and SFI (11 percent). Quebec's FSC-certified forest area is the second-largest in Canada, trailing just behind Ontario's.

Forest Company Certification Response in Quebec

As in other regions, Quebec companies certified to the ISO 14001 standard before pursuing SFM certification. A private woodlot owners' organization was the first to SFM-certify, achieving FSC certification in May 2002.[48] Further FSC and CSA certifications were delayed by difficulties particularly with regard to the overlapping tenure arrangement in the province. Louisiana Pacific certified its private and public land to the SFI standard in December 2002. Kruger and AbitibiBowater (formerly Bowater and Abitibi-Consolidated) achieved CSA certification at their various operations from 2003 to 2006.

In July 2005, Tembec became the first large industrial company in Quebec to achieve FSC certification and the first to certify to the FSC National Boreal standard. Since 2001, Tembec had been working in a partnership with the WWF to achieve innovation in SFM practices. Shortly after its FSC certification, the company signed a $120 million contract with Home Depot to supply certified forest products. Domtar FSC-certified its first forest area in Quebec in September 2005. Like Tembec, it wanted to gain international recognition and demonstrate its strong commitment to SFM. The company also certified to the FSC standard, because the standard had ENGO support and directly addressed First Nations concerns.[49]

Table 4.5

Major forest certification holders in Quebec, January 2008

Standard	Major licensees	Certified area (ha)	% of provincial certification	Date of initial certification(s)	Quebec total certified (ha)
FSC	Domtar	1,389,451	8	Sep – Dec 2005	
	Tembec	3,059,129	17	Jul 2005 – Oct 2006	
	Forestier de l'Est Lac Témiscouata	27,064	<1	May 2002	
					FSC total: 4.48 million
CSA	AbitibiBowater[1]	8,034,318	45	Nov 2003 – Dec 2006	
	Kruger[2]	2,160,335	12	Nov 2003 – Mar 2005	
	Produits Forestiers Saguenay Inc.	992,000	5.6	Dec 2005	
					CSA total: 11.19 million
SFI	AbitibiBowater	63,473	<1	Feb 2005	
	Louisiana- Pacific Canada	1,600,135	9	Dec 2002	
	Smurfit-Stone	403,251	2	Sep 2006	
					SFI total: 2 million
					Quebec total: 17.7 million

1 Abitibi-Consolidated and Bowater merged in October 2007.

2 In August 2008, Kruger announced its intention to seek FSC certification of its 2.16 million hectares of forestland in Quebec.

Source: CSFCC 2008.

The Quebec Government's Certification Response

Position on Certification

Up to the time of the Coulombe Commission's report in 2004,[50] the Quebec government adopted a neutral, passive-observer role in forest certification, viewing it as a market issue between forest companies and their customers. It intentionally did not take the lead on certification, letting each company choose whether and how to participate in the voluntary governance program. As explained by government officials, "it's up to every company to decide whether they are in or out."[51] While viewing forest certification as important to market access and useful in helping with forest management in some places, the government stressed that certification does not substitute for legislation or public decision (MRNFP 2003): "It is the people of Quebec who have to decide. Certification can add to legislation, not replace it."[52]

The government's position towards certification began to shift in 2001, with the amendments to the Forest Act (2001) and new regulations (Règlement sur les normes d'intervention dans les forêts du domaine de l'État, RNI) in 2002. For example, the MRNFP joined the CSA Technical Committee in 2002. In 2003, the Quebec government created the Commission d'étude sur la gestion de la forêt publique québécoise (the Coulombe Commission) to evaluate public forest management. During this period, Quebec companies increasingly communicated to government their difficulties in implementing certification and began to call for direct government engagement to provide greater legislative flexibility and to mandate certification (Quebec Forest Industry Council 2004). In December 2004, the independent Coulombe Commission released its report, which included eighty recommendations on the future management of Quebec's public forests.[53] The report compared Quebec with other forest producing regions and concluded that the province was lagging in several areas, including certification adoption. In particular, the commission recommended that the government mandate forest certification. In response to the report, Natural Resources Minister Pierre Corbeil explained that "we had to stop the lax approach of the past years ... we have a strong will to improve the transparency, independence and credibility of our forest management" (Forest Certification Watch 2004b).

The government recognized the importance of forest certification: that forest companies in Quebec were lagging significantly in certification adoption compared with other jurisdictions, and that government response was required. Minister Corbeil elaborated: "We realized we were behind other provinces and needed to improve this and work with the companies to solve the problems on a case-by-case basis." The government acknowledged that it needed not only to help facilitate the resolution of implementation issues but also to become engaged in order to avert future challenges, particularly by working with the certification programs to ensure that they evolved and were adapted so that they were in line with the province's forest policy.

Role in Certification Development

The Quebec government did not participate in the initial development of the CSA, FSC, or SFI standards.[54] As explained by the Montreal-based Forest Certification Watch organization, "the role of the Quebec government in terms of the development of CSA was limited, and in terms of FSC was even more discreet, resulting in significant difficulties between the requirements of the standard and the legal framework of the Quebec forestry regulations" (Forest Certification Watch 2005, 12). Recognizing how a lack of government involvement in standards development had contributed to certification implementation challenges in the province, the government subsequently joined the standard-setting and revision processes of the various certification programs. A representative from the MRNFP joined the CSA Technical Committee in 2002 and contributed to the first round of revisions to the standard. When the FSC established a branch in Quebec, the government's comfort with the FSC organization increased and the MRNFP joined the FSC National Boreal and FSC Great Lakes/St. Lawrence regional standard-setting processes as a non-voting member. The government also provided financial support for public consultation and input on the review of the draft FSC standards. It joined the provincial SFI implementation committee in 2003, a year after the committee's establishment. The government's role in the various standards development processes was to provide technical information when requested and to harmonize certification requirements with provincial forest policy as much as possible.

Role in Certification Implementation

During the 1990s, the Quebec government took a position of non-interference in certification implementation, focusing instead on policy reform and leaving certification to the market. Companies ran into trouble, however. They had two main challenges: shared volume tenures and legislative alignment. First, it was hard for them to achieve agreement among all of the overlapping licensees in the shared forest area. As one company explained, "it's not difficult to get agreement on the law but with certification it's different. It's hard to get visions to align."[55] Second, certification requirements that went beyond the law were not always consistent with the law. One interviewee explained: "The FSC Boreal standard calls for a 60m riparian buffer and Quebec legislation requires 20m. What should a company do – break the law?"[56] Companies were stalled trying to figure out how to proceed without suffering legal penalties. They could apply for an exception under clause 25.3 in the Forest Act but it could take up to eighteen months for approval.

Companies and other stakeholders in Quebec were enthusiastic about certification but frustrated with the implementation hurdles, and were looking for government commitment and engagement to support and facilitate certification. In response, in June 2004, the government participated in a series of meetings with the industry to learn about company efforts and to help resolve the certification implementation issues.[57] In early 2005, it also turned its attention to addressing the various recommendations included in the Coulombe Commission report, including recommendation 7.16, regarding greater government engagement in certification: "that the Department adopt a proactive forest certification approach, notably in the following areas: promoting and actively supporting territorial certification; seeing that 3rd party participation processes for planning forest management activities be recognized by the certification system to avoid duplications; and participating more actively in the development and improvement of forest certification systems."

Role in Certification Enforcement

A key part of the Coulombe Commission report was recommendation 7.15, regarding certification enforcement. It recommended that "all forest management units in Quebec public forests be engaged in a forest certification process under an internationally recognized standard by the end of 2007." All forest companies in Quebec had made submissions to the commission advocating greater legislative flexibility to meet certification requirements. On 15 April 2004, the Quebec Forest Industry Council (QFIC) had formally called upon the provincial government to "require all companies charged with planning and carrying out forestry work on public forest lands to have their practices certified by an independent accredited agency" (Quebec Forest Industry Council 2004). The QFIC wanted to stimulate certification in the province and assure the public and customers about the sustainability of Quebec forest practices. It explained in its media release that mandatory certification on public land would "assure the transparency, neutrality and credibility of all dimensions of Quebec's forest system ... also producing positive impacts not only on export markets for its products but also vis-à-vis citizens concerned about the sustainability of the resource."

A year later, in April 2005, the government was still examining the issue and indicated that it had no intention of legislating certification: "We are in favour of certification but it is not an obligation. The government's role is to help solve certification problems."[58] Two years later, however, after considering the Coulombe Commission recommendations, the government signalled its intention to enable the Natural Resources Minister to mandate certification. Bill 39, adopted in December 2007 by the National Assembly, granted the minister the power, if desired, to require that agreement holders obtain forest certification from an independent agency with SFM standards applicable to Quebec's forests (MRNF 2008, 62).

Ontario

Ontario is Canada's most populated region and has the largest and most diversified provincial economy. It is also the country's third-largest forested and timber-producing province, relying on the vast northern boreal forest to support its forest sector. Although 90 percent of the population lives in the southern part of the province, there have been tensions in balancing industrial, ecological, and social forest values. As explained by the Canadian Parks and Wilderness Society (CPAWS), the fundamental challenge is that the boreal region is the most important to the forest industry but is also the most intact forest region in the world, and therefore essential to protect.

The Ontario government has been a certification leader, directly engaging in certification development, implementation, and enforcement. The government's enthusiasm was demonstrated early on, when it announced in April 2001 its intention to certify the entire province to the FSC standard. Although this commitment was immediately retracted, the province continued to directly engage in certification as a means of promoting alignment between its comprehensive forest legislation and SFM certification requirements. In 2004, the government mandated certification on public land across the province.

The Provincial Context

Ontario has the greatest forest coverage of any province in Canada. Two-thirds of its 70 million hectares of forestland is boreal forest.[59] The remainder consists of deciduous hardwood forest in the southern part of the province and mixed forest in the Great Lakes–St. Lawrence region. Black spruce is the dominant tree species, followed by poplar and jack pine. Because there is such a large amount of boreal forest, Ontario has attracted the attention of environmental organizations that want to protect this vast intact forest region.

Ontario's forests are 89 percent publicly owned. The Forest Division of the Ontario Ministry of Natural Resources (OMNR) is responsible for regulating management of the public forests through Sustainable Forest Licences (SFLs) under the Crown Forest Sustainability Act (CFSA) established in 1994. The SFLs are twenty-year, area-based tenure licence agreements that grant cutting rights to forest companies in a specific Crown forest area – a forest management unit. In return, licensees are responsible for conducting forest management planning, inventory, monitoring, and reforestation, and for complying with all provincial forest laws. The OMNR reviews and renews licences every five years. The industry licensees source 75-80 percent of their fibre from Crown lands.

Eleven percent of Ontario's forests are privately owned through various private tenure arrangements, ranging from cottage properties to large industrial holdings in northern Ontario. Private forests account for over half of the hardwood forest harvested in the province and are supported by the Ontario Stewardship Program (OSP). Through a network of forty community-based councils, the OSP provides information and expertise to private landowners to encourage sustainable forest management practices.

Ontario is a major wood and paper producer and exports most of its products to the United States. The province's main forest product exports are softwood lumber, newsprint, and wood pulp. Like all other Canadian forest producing provinces, Ontario has faced major challenges with respect to the economic sustainability of its forest sector.[60]

Forest Certification Uptake in Ontario

Ontario has 26.3 million hectares of certified forest and the greatest amount of FSC-certified forest of any province. Ontario accounts for 17 percent of Canada's forests, 20 percent of Canada's total certified forest, and approximately 50 percent of the country's FSC-certified forest. Within the province, the certified forest is divided between the various standards – 42 percent are FSC-certified, 28 percent CSA-certified, and 30 percent SFI-certified (Table 4.6). Ontario was an early adopter of certification; the first FSC certification in Canada was in the province's Haliburton Forest and Wildlife Reserve in March 1998. The major licensees certified to the FSC and CSA programs for the most part before 2004, whereas SFI certification occurred from 2005 to 2007.

Forest Company Certification Response in Ontario

As in other provinces, large industrial licensees such as Abitibi Consolidated, Bowater, Tembec, Weyerhaeuser, and Domtar certified to the ISO 14001 standard prior to SFM certification. Unlike in other provinces, however, the FSC had a much better reception in Ontario. As mentioned above, Haliburton Forest was Canada's first FSC adopter, certifying its small 22,000-hectare privately owned forest in March 1998.[61] The Eastern Ontario Model Forest was also a certification leader in the province.[62] It began looking at FSC certification in early 1999 and achieved FSC resource manager certification in 2003.

Tembec and Domtar have been the sustainability leaders among industrial forest companies in Ontario. Sensing increasing conflict in the province's boreal forest, Tembec encouraged other industry players, including Domtar, to form partnerships and working arrangements with environmental organizations, First Nations, and government. The result was the province's Living Legacy Policy and the achievement of FSC certification by each company. In 2000, Domtar achieved its first FSC certification in Canada, in the Gilmour

Table 4.6

Major forest certification holders in Ontario, January 2008

Standard	Major licensees	Certified area (ha)	% of provincial certification	Date of initial certification(s)	Ontario total certified (ha)
FSC	Algoma Forest	951,004	3.6	Jun 2005	
	Domtar	3,471,088	13	Apr 2000 – Aug 2007	
	Haliburton Forest and Wildlife Reserve	21,998	<1	Mar 1998	
	Nipissing Forest Management	1,147,501	4	May 2003	
	Tembec	3,969,578	15	Apr 2003 – Jan 2006	
	Vermillion Forest Management	648,897	2.5	May 2006	
	Westwind Forest Stewardship	855,446	3	Feb 2002	
					FSC total: 11.1 million
CSA	AbitibiBowater	4,617,384	17	Dec 2002 – Feb 2005	
	Domtar	1,760,000	6.7	Dec 2003	
	Weyerhaeuser	1,016,000	3.8	Apr 2005	
					CSA total: 7.4 million
SFI	AbitibiBowater	3,316,892	12.6	Jan 2005	
	Long Lake Forest Products Inc.	746,484	2.8	Mar 2007	
	McKenzie Forest Products Inc.	721,540	2.7	June 2007	
	Terrace Bay Pulp	1,927,336	7.3	Jan 2005	
					SFI total: 7.8 million
					Ontario total: 26.3 million

Source: CSFCC 2008.

Forest near the city of Trenton. Shortly after, Tembec's April 2003 certification of the Gordon Cosens forest in Northern Ontario (over 2 million hectares) was the largest FSC certification in North America and the first boreal certification in Canada. Domtar and Tembec together account for close to 35 percent of Ontario's certified forest area.

Other large integrated forest companies such as Abitibi-Consolidated, Bowater, and Weyerhaeuser were FPAC members and committed to SFM-certify all of their forestlands across Canada to meet the FPAC membership requirement. By 2005, these companies had achieved CSA certification for their Ontario forestlands. Since 2005, several licensees have certified to the SFI standard, including AbitibiBowater (Abitibi Consolidated and Bowater merged in 2007) on its private land as well as Terrace Bay Pulp, Long Lake Forest Products, and McKenzie Forest Products.

Compared with companies in other provinces, Ontario forest companies encountered few issues or challenges in achieving forest certification. Ontario forest legislation aligned well with certification requirements and companies were familiar with third-party audits since the Crown Forest Sustainability Act included an independent forest audit process.

The Ontario Government's Certification Response

Position on Certification

Before 2000, the Ontario government's role in certification was largely passive – observing and learning about industry response, conducting comparisons of the different standards, and assessing how certification related to the province's forest regulatory regime. During this period, the Ministry of Natural Resources activity largely focused on revising the forest legislation as well as completing a timber class environmental assessment for the province. The CFSA was approved in 1994 and the Environmental Assessment Act (EAA) amended in 1996. The CFSA introduced significant changes to the provincial forest tenure system to meet increasing public expectations regarding ecosystem-based forest values. Regulatory reforms continued in 1997 with the Lands for Life provincial land-use public consultation and the release of the Living Legacy land-use strategy in 1999. The Ontario Forest Accord was signed in March 1999 (OMNR 1999). The accord was a consensus agreement between environmental organizations, the forest industry, and the provincial government to balance forest protection with timber production values. Its Recommendation 23 stated that the parties would encourage and support international forest certification activities undertaken by companies.

With the signing of the accord, the government began to take a more active interest and direct role in certification. While regulatory reforms continued, the government was also highly confident that the province's

SFM regulations and unprecedented forest policy consensus far exceeded certification requirements. On 23 March 2001, it therefore issued an unexpected press release. Together with FSC International, the Minister of Natural Resources announced an agreement whereby Ontario's legislative requirements would be formally recognized by the FSC, effectively FSC-certifying all Crown land in the province.[63] The joint announcement declared that "Ontario companies are already engaged in the practice of sustainable forestry under the province's stringent forestry laws ... FSC will tell the world that the Ontario government has worked with all stakeholders to ensure that our standards are met." While the FSC and government press releases were identical in content, the OMNR release included its own title, proclaiming, "Ontario first in the world to receive environmental forest certification" (OMNR 2001).

Reaction to the OMNR/FSC announcement was swift and overwhelmingly negative. Industry and ENGOs argued that the pronouncement was at least premature, as no formal assessment of Ontario's laws and policies had taken place. The announcement was also deemed to undermine the FSC regional standard-setting process that was underway in the province. The Sierra Club of Canada exclaimed that the press release contained "egregiously misleading and unsubstantiated statements regarding the sustainability of Ontario's forest practices."[64] WWF-Canada pointed out that making advance certification claims was not permitted by the FSC; that only accredited certifiers could grant certification, and not FSC International; and that the FSC was not able to blanket-certify an entire jurisdiction – that the appropriate mechanism for certification was through the development and approval of regional standards.[65]

The FSC removed the announcement from its website on 30 March 2001 and issued a letter of clarification on 4 April 2001. The letter addressed the various concerns and gave reassurances that the announcement did not mean an advance approval of Ontario's practices but rather a commitment between OMNR and the FSC to work towards a more formal agreement "whereby Ontario's forests could become eligible for FSC certification, but we are not there yet" (FSC 2001). The exclusive FSC certification of the provincial forest management regime did not proceed, but the government continued to play a proactive role in promoting *all* certification systems and by 2004 mandated certification (to the CSA, FSC, or SFI standards) on Crown land across the province. Its position was to encourage companies to seek certification by an accepted independent third-party organization, and included promoting certification as a tool that forestry operations could use to have their forest management practices assessed in order to maintain access to consumer markets. The OMNR described (on its website) its certification role as "providing technical and policy advice both during the development of certification systems and to forest companies seeking certification of forest lands in Ontario."

Role in Certification Development

The Ontario government participated in the development of the CSA standard and took a particular interest in the FSC process. A representative from the OMNR served on the CSA Technical Committee. Although the ministry was not an official voting member, OMNR staff requested and were permitted to attend both the FSC Great Lakes/St. Lawrence and FSC Ontario Boreal regional standard meetings.[66] They attended every meeting to provide technical information and guidance as requested and to ensure the alignment of FSC requirements with the provincial forest policy and regulations. An OMNR staff person was also a member of the FSC National Boreal Coordinating Committee.[67] The ministry explained: "We asked to participate on the FSC regional committees and were allowed to attend as non-voting observers. We weren't allowed to speak until the end but the committees started to use the MNR for scientific and technical advice. For example, we had experts on hardwoods and they used our technical documents."[68] The government's key interest in engaging in both the CSA and the FSC standard development processes was to ensure consistency with Ontario's forest legislation, specifically, the province's forest management planning criteria and indicators and its compliance audit protocols.

Role in Certification Implementation

Because it was actively engaged in the development of the various standards, the government knew that the province's forest legislation and regulations aligned well with certification requirements.[69] It was therefore keen to help the Ontario forest industry gain international recognition for the industry's sustainable forestry practices. The Minister of Natural Resources explained that "certification tells the world that we are among the leaders in managing our Crown forests responsibly ... Ontario's forest legislation and policy framework provide a strong foundation for those seeking forest certification in Ontario" (OMNR 2002b).

Having played a supporting role in the development of the CSA and FSC standards, the OMNR next assumed a direct and active role in facilitating certification implementation, including initiating ongoing dialogue with the industry to learn of certification challenges, holding certification training sessions, creating guidebooks, offering scientific and technical advice, making staff available during certification audits to answer questions on forest policy and to provide supporting compliance audit evidence, participating in certification pilot projects, and assessing how to streamline certification and compliance audits.

Overall, the forest authorities worked with the various certification programs to simplify controls and avoid duplication in order to facilitate certification efforts. For example, in November 2002, the OMNR became the first government agency in Canada to establish a formal forest certification memorandum of understanding with the Standards Council of Canada (SCC). The MOU recognized the similarities between the CSA requirements and the OMNR's regulatory requirements and committed both parties to work to facilitate CSA certification in the province. As outlined in the OMNR press release, "the MOU between OMNR and SCC allows each to recognize the other's requirements and through a commitment to co-operate, arrive at more efficient processes leading to registration or certification" (OMNR 2002a). As a follow-up to the MOU, the OMNR conducted a gap analysis comparing CSA requirements with provincial forest policy, and developed guidebooks to help forest companies implement and achieve CSA certification.

The OMNR also developed a collaborative action plan with FSC Canada to address the FSC certification system in Ontario (OMNR 2006a). Key action items included identifying common approaches to meeting FSC requirements, reviewing existing FSC certifications to address any provincial barriers to certification, and identifying and reducing redundancies in audit requirements. In 2006, the Ontario government and FSC Canada agreed to compare the FSC National Boreal standard with the province's independent forest audit (IFA) requirements.

Role in Certification Enforcement

On 1 April 2004, Ontario Minister of Natural Resources David Ramsay announced that all SFL holders would be required to certify to an accepted performance standard (FSC, CSA, or SFI) by the end of 2007. The government expected the requirement to achieve three results: gain market recognition for Ontario's forest products, confirm Ontario's high-quality legislative and regulatory framework by independently verifying SFM practices in the province, and accelerate the certification of Ontario's forests to ensure that the Ontario forest industry remained competitive with neighbouring jurisdictions. Ramsay explained the important role of certification in supplementing provincial forest policy: "The government is making real, positive change with its plan to build on the existing regulatory requirements that must be met in order to undertake forest operations in Ontario."

At the time of Ramsay's announcement, 24 percent of the managed forestland in the province was certified. The OMNR initiated and maintained a dialogue with the industry in order to understand and address the challenges of meeting the 2007 mandatory requirement. By September 2006, the proportion of certified Crown forest had increased to 60 percent. In May 2007, the regulations were changed to permit the minister to make certification mandatory through an amendment to an existing Sustainable Forest Licence.[70]

As part of the government's overall effort to improve the efficiency and effectiveness of provincial forest management requirements and practices, the OMNR also supported efforts to align certification enforcement audits with the province's independent forest audit process. For example, in 2002 the government participated in a pilot on the Crossroute Forest to test the feasibility of aligning the CSA and IFA processes. The OMNR also encouraged companies to pilot test streamlining the two audit processes. Two company trials were conducted: the 2005 FSC audit of the Spanish Forest and the 2006 FSC annual audit of the Gordon Cosens Forest. The government's Forest Process Streamlining Task Force highlighted the issue in 2006 with its recommendation that the OMNR "develop a policy and procedure for the integration of third-party certification management systems into the forest operations compliance program" (OMNR 2006b, 34). In December 2006, Senes Consultants completed a five-year review of the government's IFA program. It concluded that although the IFA and certification processes had been essentially parallel and completely separate up to that point, there was increasing overlap and there was an opportunity to integrate the two programs, particularly for annual certification audits to leverage IFA evidence and results (Senes Consultants 2006, 34).

Overall, the Ontario government directly engaged in certification as a means of demonstrating the strength of its forest policy and further inspiring and advancing sustainable forest management. In particular, its objective in co-regulating forest certification was to ensure alignment and avoid conflicts and redundancies between certification and provincial forest policy, as well as encourage continual learning and relationship building among stakeholders.

Provincial Government Engagement in Certification

As outlined in the previous sections, Canadian provincial governments' engagement in certification ranged from indirect to direct co-regulatory approaches at the development, implementation, and enforcement stages. Figure 4.4 summarizes this variation across a spectrum of intervention, from observing, cooperating, enabling, and endorsing certification systems to directly mandating certification. The figure also indicates the shift in provincial government response towards increasingly direct certification approaches as certification systems gained legitimacy through forest industry uptake, and as governments' knowledge of certification increased.

Standards Development

All provinces cooperated in the development of the various certification standards. For example, each provincial government had representatives

Figure 4.4

Provincial government responses to certification

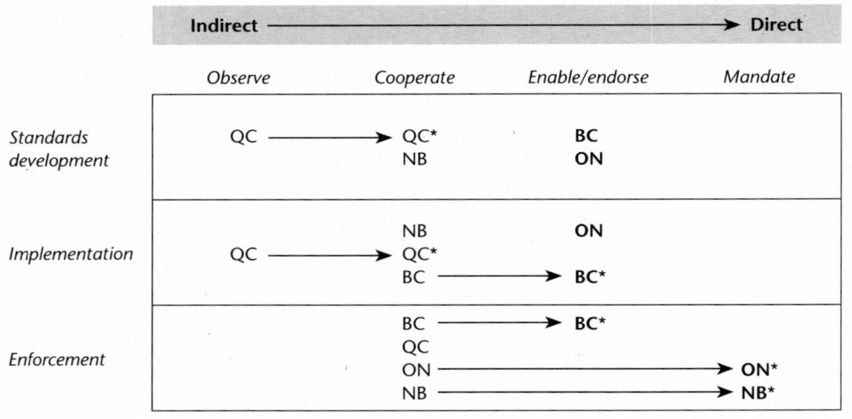

* Highlights a shift in provincial government approach to certification over time.

who served on the CSA Technical Committee, participated in the respective FSC regional standard-setting processes as non-voting observers, and served on provincial SFI implementation committees. There was variation, however, in the extent to which the provinces engaged in standards development and also in the timing of participation. Specifically, the Ontario and British Columbia governments played a direct role in enabling standards development, whereas the Quebec government lagged in its participation. British Columbia encouraged the development of a national SFM standard and promoted it to customers and offshore markets, and the Ontario government directly endorsed the FSC standard in 2001 by announcing its partnership with FSC International to certify all provincial forestland managed under the Crown Forest Sustainability Act. In contrast, the Quebec government joined the CSA Technical Committee later in the process, for the 2002 revisions to the CAN/CSA-Z809 standard rather than for its initial development and its publication in 1996. Also, Quebec cooperated with the FSC only after an FSC office had been established in Montreal. The evolution of the Quebec government's co-regulatory role in certification standards development is shown in Figure 4.4 by the shift QC → QC*.

Certification Implementation

The implementation stage followed a similar pattern. At a minimum, all provinces cooperated in helping forest owners achieve certification, but

Ontario stood out in taking an early and direct co-regulatory role in enabling certification implementation whereas Quebec lagged in its engagement in certification. British Columbia and New Brunswick were supportive of industry certification implementation efforts and provided assistance as requested.

The Ontario government directly engaged in certification implementation by establishing cooperative agreements with the CSA and FSC standard-setting bodies. The OMNR also proactively provided certification training and information to licensees and prepared guidebooks explaining the certification/regulatory alignment to enable certification across the province. The Quebec government adopted a hands-off approach to certification implementation, viewing it as a market issue between companies and their customers, but it gradually shifted to a more cooperative role as its awareness of certification increased (QC → QC*).

Although the BC government recognized the importance of facilitating certification implementation and created a multi-stakeholder advisory council in 2000 to study and recommend government options, it was not until 2004, with the introduction of the new results-based forest legislation, that its role shifted towards directly enabling certification (BC → BC*). The Forest and Range Practices Act enabled certification as it encouraged greater private governance initiative. Under this act, industry became responsible for meeting SFM outcomes (rather than prescriptive operational requirements), and certification became an important vehicle for forest operator innovations and for tracking SFM improvements to achieve and demonstrate legislative compliance. In 2004, the BC Forest Ministry commissioned studies to evaluate the alignment of certification and provincial forest policy in order to lessen the province's regulatory burden and improve the efficiency of certification implementation.

Certification Enforcement

Initially, all governments largely observed certification enforcement and participated on an as-requested basis in certification audits by providing clarification of provincial forest policy and regulations. The BC, Ontario, and New Brunswick governments also had an interest in streamlining the public and private governance processes, and undertook studies to evaluate the possible integration of certification and provincial compliance audits. Both Ontario and New Brunswick mandated certification, making it a requirement for all licensees operating on Crown land (ON → ON*; NB → NB*). The BC government had no interest in mandating certification, but it did directly endorse certification enforcement by facilitating and promoting certification of the BC Timber Sales program (BC → BC*). As in BC, the Quebec government had no interest in establishing a legislative requirement for certification and took the position that certification should be left entirely

Table 4.7

The role of provincial governments in certification enforcement, 2005

Province	Certification requirements
British Columbia	Considering SFM-certifying BC Timber Sales.
Ontario	On 1 April 2004, issued an announcement requiring all major licensees to certify by 2007.
Quebec	Considering Coulombe Commission and Quebec Forest Industry Council recommendations to mandate certification by 2007.
New Brunswick	All licensees required to ISO-certify by December 2002 and SFM-certify by December 2003.

to market forces. With industry making direct requests for the government to mandate certification, however, and with the Coulombe Commission making similar recommendations, the province amended legislation (in December 2007) to permit the Natural Resources Minister to require certification on public forestland.[71] Table 4.7 summarizes the role of provincial governments in certification enforcement as of 2005.

It is evident that there have been a range of provincial government co-regulatory roles at the development, implementation, and enforcement stages of certification, including instances of direct enrollment when governments certified their own forest programs (BC) or legislated certification (New Brunswick and Ontario). Governments also played a direct role in enabling certification by dedicating public resources to standards development (Ontario, BC) and implementation (Ontario), and through the introduction of legislative amendments to facilitate the policy role of private governance initiative (all provinces).

Factors Influencing Certification Co-Regulation

Provincial governments' engagement in certification has not been based on happenstance, nor has it been unstable with the election of different political parties to power. Rather, government responses have been intentional and consistent across electoral cycles with the goal of ensuring the optimal co-governance of the provinces' forests and forest sector. Various conditions and factors influenced provincial government certification co-regulation.

Background Conditions and Provincial Governments' Certification Rationale

As outlined earlier, the provinces examined in this chapter have very similar forestry regimes. Most public forestland is under provincial government authority, all provinces delegate public forest management responsibilities largely to industrial forest companies, and all export the vast majority of their forest products to US markets. In addition, all of the provinces have faced societal and market pressures to balance forest values and maintain global economic competitiveness. These conditions contributed to each provincial government's rationale for engaging in certification.

Public Forestland Ownership and Responsibility

As major forestland owners, provincial governments engaged in certification to enhance public forestland management and facilitate continued market access for public timber. As the principal forest legislators in the province, they also sought to protect their forest agenda and ensure the alignment of public and private forest policies. Governments also had an interest in leveraging certification (for example, the audit processes) to streamline regulatory enforcement and reduce their regulatory costs.

Delegated Industrial Forest Management

Canadian forest policy regimes are typically described as public forest management for private timber production. All provincial governments have established tenure arrangements that delegate responsibility for public forestland management to industrial forest licensees. Engaging in certification thus presented governments with an opportunity to increase the transparency of their delegated forest management tenure arrangements.

Forest Sector Export Dependency and Declining Global Competitiveness

Canadian provinces are dependent on the export of their forest products, particularly to US markets, and are all under pressure to maintain competitiveness in global markets. Over the past decade, all provinces faced reduced provincial budgets and declining forest sector profitability and royalties, which caused them to seek ways to reduce costs and stimulate their regional forest economies. Engaging in certification was therefore a way to ensure market access for provincial forest products, gain market advantage for provincial forest producers, and enhance the province's long-term competitiveness.

SFM Policy Frameworks

Beginning in the 1980s, provincial governments began to develop and implement SFM policy frameworks in response to increasing societal expectations to protect non-timber values, and to the availability of better scientific information about the ecological functioning of forests. This meant introducing a policy agenda and formulating policy that, among other things, balanced forest values and enhanced public participation in forest management decisions. The SFM policy frameworks were guided by SFM criteria and indicators developed through international processes and the Canadian Council of Forest Ministers committee and also included in Canada's National Forest Strategy. Thus, an additional provincial government rationale for engaging in certification was not as a substitute for inadequate SFM policy but rather to leverage an additional policy vehicle that would complement their efforts to implement and track SFM criteria and indicators and engage local stakeholders in forest planning and decision making.

Factors Influencing the Variation in Government Certification Role

It is difficult to isolate a particular driver that led each provincial government to respond to certification the way it did. Rather, various factors interacted with each other and with the background social, political, economic, and environmental conditions to influence a government's response. Three factors are particularly important in explaining not just *why* provincial governments engaged in certification but also *how* they responded to it: industry expectations of government's role, ENGO advocacy, and alignment between certification and policy depending on the stage of the provincial forest policy cycle. These factors were common to all regions, yet played out differently in each in terms of their interaction and influence on government certification response.

Industry Expectations of Government's Role

The forest industry in Canada has historically held significant power and policy influence through close and privileged access to government decision makers. Forestry is a key sector of the Canadian economy and supports resource-based communities across the country. Governments depend on a viable forest sector for economic growth, employment, and prosperity. Provincial governments have also established close negotiated alliances with their respective forest companies for the management of a large percentage of the Crown forest. Overall, Canadian provincial forest regimes are typically characterized as "clientelist" in nature, with state and business members traditionally dominating provincial forest policy networks.[72] Given the privileged position of the forest sector in Canada, forest company expectations of government's role can be expected to have an important bearing on how governments respond to certification.

Beyond the historical pattern of forest industry influence, certification has created a new business/government policy dynamic that reinforces the importance of evaluating industry expectations of government. Traditionally, industry behaves as an interested group, lobbying or "pushing" to influence government decision-making authority. Certification has introduced a new "pull" dynamic, however. As outlined in Chapters 2 and 3, with certification private governance systems, corporate actors have gained private agenda-setting and rule-making authority within an expanded political space that goes beyond traditional government authority. This has created a tension between public and private authority, resulting in a *pull* on governments to respond to industry CSR to prevent erosion of their policy sovereignty. A strong industry position either encouraging or discouraging government engagement in certification is therefore influential in shaping how a government approaches the co-regulation of certification.

The interviews I conducted in 2004-05 with forest managers and executives of forest companies across the four provinces revealed a range of corporate perspectives and expectations about government's role in forest certification. Although all of the interviewees were speaking for themselves (in their company role) as opposed to necessarily conveying a formal company position, all stated that a critical role of the provincial government was to provide a clear SFM legal framework that would enable certification and also provide sufficient flexibility for companies to go beyond the law if necessary to meet certification requirements. All interviewees supported a government role in standards development to ensure policy/certification alignment, and also supported government cooperation in facilitating certification implementation. Perspectives were divided, however, as to whether government should mandate certification. Company interviewees in New Brunswick and Quebec were unanimously supportive whereas those in BC were all strongly opposed to government legislation of certification. Companies in Ontario were neutral to unsupportive.

New Brunswick companies were supportive as they felt that mandatory certification would position the provincial forest industry well in the market and send their customers a positive message about forest practices in the province. Quebec companies felt that mandating certification would ensure important third-party oversight so that the public and customers could verify SFM practices in Quebec. Ontario companies had different perspectives but were generally neutral to unsupportive of the government's announcement to mandate certification, noting that it would take away industry's ability to "walk away" if certification requests became unreasonable. Finally, companies in BC adamantly opposed government mandating of certification as the province was emerging from a heavily prescriptive period under the Forest Practices Code and companies were happy with the new deregulatory environment under the Forest and Range Practices Act.

Overall, industry expectations regarding the role of government in certification enforcement (whether or not to mandate certification) were aligned with governments' certification response in New Brunswick and British Columbia but not in Ontario or Quebec (Table 4.8).

As Figure 4.5 shows, the New Brunswick and BC forest industries were influential in shaping government response largely because, within each province, companies took similar approaches to certification and held strong, consistent positions on the role they wanted government to play in forest

Table 4.8

Industry and government alignment regarding the mandating of certification, 2005

Province	Industry attitude towards provincial mandating of certification	Government response
New Brunswick	Yes	Mandated certification in April 2002
Ontario	Neutral/No	Mandated certification in April 2004
British Columbia	No	Not mandated
Quebec	Yes	Not mandated

Figure 4.5

Influence of industry expectations on certification enforcement
Here "alignment" refers to the match between government certification response and industry expectation.

Government response	Company expectations of government	
	Mandate	*Don't mandate*
Mandated	***New Brunswick*** Alignment encouraged by strong and uniform company expectations.	***Ontario*** Misalignment influenced by heterogeneous company expectations.
Not mandated	***Quebec*** Misalignment influenced by a lag in government certification awareness.	***British Columbia*** Alignment encouraged by strong and uniform company expectations.

certification enforcement. Forest companies in Ontario were less influential because companies took different approaches to certification (whether to certify to the FSC, CSA, or SFI standard) and the Ontario forest industry as a whole was heterogeneous in its expectations of government and did not take a particularly strong position one way or another towards the mandating of certification. Quebec forest companies also took different approaches to certification but, unlike in Ontario, were united and strong in their demand for greater government engagement, including the mandating of certification. At the time of the survey, however, the Quebec government lagged behind the other provinces in terms of certification awareness and was not yet prepared to respond.

In summary, industry expectations of government had varying influence across the provinces in terms of shaping how governments responded to certification, particularly at the enforcement stage. The extent to which forest companies were consistent in their certification approach and were strong and united in their position towards the government's role contributed to how much influence they wielded.

ENGO Advocacy

Over the past several decades, there have been increasing societal pressures on governments to address environmental forest values. Local, national, and transnational environmental NGOs (ENGOs) launched campaigns to encourage governments to conserve and protect threatened environments (old-growth forests, species at risk, biodiversity and wildlife habitats) and implement sustainable forest management regimes that balanced environmental, social, and economic forest values. Besides traditional lobbying, ENGOs also adopted a new advocacy approach – market campaigns to specifically target the large buyers of forest products, to reduce the demand for chlorine-bleached paper, encourage recycled content, and discourage the sourcing of tropical and old-growth timber. To varying degrees, all provincial governments were faced with both the traditional and the new market-based forms of ENGO lobbying pressure. From the 1980s onward, they faced the challenge not only of responding to ENGO-led criticism of forest policy but also of ensuring that their forest sectors were not unfairly affected by adverse trade decisions resulting from the market campaigns.

Certification grew out of the ENGO market campaigns and offered governments a unique co-regulatory opportunity. First, certification could address the market pressure and possibly secure long-term global market access for provincial forest products. And second, certification would involve independent, third-party assessments of provincial forest practices that could, if successful, demonstrate the consistency of provincial forest policy with international SFM principles, justify the province's overall forest management policy approach to skeptical ENGOs, and secure market and public trust.

ENGO forest policy advocacy and market campaigns not only provided a rationale for government engagement in certification but also influenced how governments responded to it. In particular, specific advocacy issues and whether the lobbying campaign was local as opposed to globally driven affected government's role in certification. These factors varied between provinces, contributing to the variation in government co-regulatory certification response.

For example, compared with the other provinces, British Columbia faced a very strong global ENGO lobby made up of groups such as the Rainforest Action Network out of San Francisco and Greenpeace Germany. This led the government to adopt a hands-off approach towards certification implementation and enforcement and to focus instead on promoting the adequacy of forest legislation and land-use planning processes and on guarding the provincial forest agenda from the influence of these "outside" groups. The specific issues of the ENGO campaigns (protection of coastal old-growth forests, halting of clearcutting forestry practices, and so on) shaped the BC government's certification response. The ENGO advocacy issues were not easily addressed without significant restrictions on the provincial timber supply, and therefore significant adverse impacts on the provincial economy. The government feared that the FSC-BC regional standard would adopt similarly challenging requirements, so it not only participated enthusiastically in the CSA standard development process but also took an active role in promoting the Canadian standard to offshore markets. Overall, in response to ENGO advocacy, the BC government took an approach to certification that ensured that forest companies in the province had a viable certification option and that the provincial forest policy agenda remained under the government's sovereign authority.

As opposed to direct pressure, an *expectation* of global ENGO advocacy influenced the Ontario government's response to certification. The government had experience with ENGO protests in the late 1980s over old-growth logging in the northern Temagami region of the province, and knew that it had reached an unprecedented agreement with local ENGOs (the Ontario Forest Accord) in 1999. So when it sensed a mounting global ENGO campaign targeting Ontario's northern boreal forests, the government sought to head off the campaign by announcing in 2001 a partnership agreement with FSC International to recognize Ontario's legislative requirements and essentially FSC-certify all public forestland in the province. When FSC provincial certification proved unfeasible, the government adopted its direct co-regulatory approach of mandating forest certification to any of the recognized SFM standards.

In Quebec, the *absence* of a strong local or international ENGO forestry lobby contributed to the government's initial "backseat" approach to certification and subsequent lagging certification response. In the early 1990s, the government was not under significant public or ENGO pressure to justify its SFM practices. This changed in 1998, with the release of the film *L'Erreur boréale*.[73] The film conveyed images of destructive forestry practices in Quebec's publicly owned boreal forest, and this sparked public awareness and the beginning of ENGO boreal campaigns in the province. Heightened advocacy increased the Quebec government's awareness of certification and spurred its decision to join the CSA standard revision process and the FSC regional standard development process already underway.

The *history* of controversy surrounding spruce budworm spraying influenced the New Brunswick government's decision to mandate forest certification.[74] Aerial spraying for spruce budworm began in the province in 1952 (and continues to this day). Although public protests and ENGO campaigns, combined with scientific evidence of deleterious health and environmental effects, resulted in some success in encouraging the substitution of less harmful insecticides, the battle over spraying in New Brunswick was largely won by powerful economic interests (May and Rogers 1982; Sanberg and Clancy 2002). Forestry is critical to the provincial economy, and government and the industry argued that annual spraying was essential to protecting the provincial timber supply, and hence fundamental to securing the province's economic future. Thus, when J.D. Irving dropped its FSC certification mainly because it disapproved of the FSC Maritimes regional standard's requirement to significantly reduce biocide spraying, the government supported its concerns and advised the industry to seek an alternative, feasible certification standard. The government then mandated certification to repair any damage to market confidence in New Brunswick forest practices that may have resulted from the Irving/FSC affair.

The Forest Policy Cycle and Certification/Policy Alignment

Whether a government was in the early or late stages of its policy cycle (namely, agenda setting and policy formulation versus implementation and evaluation of established policy) had an effect on its co-regulatory response to certification, which was largely based on whether the government perceived the certification standards as complementing or competing with the policy being formulated or with established policy. For example, when certification emerged in the early 1990s, Ontario and BC were both in the early stages of their periodic forest policy cycle, and both provinces eagerly engaged in certification standards development. New Brunswick and Quebec were at a later stage and played a less direct initial role.

For Ontario and BC, it was crucial to ensure certification alignment with the newly formulated provincial forest agenda and forest legislation. The two provinces diverged, however, in their approaches to certification implementation and enforcement. BC assumed a passive role whereas Ontario directly enabled implementation and mandated certification. Ontario embraced certification as a complement to and means of promoting its new Forest Act, whereas the BC government perceived certification as potentially competing with the newly established provincial Forest Practices Code and was therefore determined to maintain its policy authority. By the start of its next policy cycle and the introduction of the results-based Forest and Range Practices Act, the BC government was more confident of certification alignment with provincial forest policy, and consequently committed to certifying the province's Timber Sales program.

When certification emerged in the early 1990s, both New Brunswick and Quebec were in the later stages of their forest policy cycle; both had well-established Forest Acts that had been formulated and implemented in the 1980s. The two governments responded differently to certification, however. New Brunswick assumed a fairly passive role, simply cooperating in standards development and implementation but taking on a direct role in mandating certification. The New Brunswick government directly co-regulated certification enforcement in the province as it was confident that its Forest Act (1982) aligned well with certification requirements (CSA and SFI, though not FSC), given the rapid certification uptake in the province. It also wanted to assert its leadership in forestry policy in Canada.

Quebec also had an established Forest Act (1986, revised in 2001), but because it was opposed to directly co-regulating certification, it largely ignored the matter, leaving it to market forces. The government believed that amendments made to the forest legislation during the 1990s were sufficient to achieve provincial forestry objectives. When it entered the formulation stage of its policy cycle in early 2000, it began to recognize the importance of engaging in the certification standards-setting process to ensure certification/policy alignment, and it subsequently joined the CSA Technical Committee. Industry then began expressing its concerns that there were certification challenges in the province largely stemming from insufficient legislative flexibility to meet certification requirements. With the release of the Coulombe Commission report in 2004, the government entered the agenda-setting stage of its forest policy cycle and became more engaged in facilitating certification implementation; it also introduced amendments to the legislation to enable it to enforce certification.

As summarized in Figure 4.6, provincial governments in the early stages of their policy cycle had an incentive to directly engage in certification standards development in order to ensure policy alignment (Ontario and BC). If a government believed that certification aligned well with its provincial forest policy, it was more likely to directly engage in the co-regulation of certification implementation and enforcement as a means of reinforcing and supplementing forest policy (New Brunswick and Ontario). If there were concerns, however, that certification requirements did not align particularly well with provincial forest policy, a government was more likely to assume a more passive, indirect role and actively promote the sufficiency of provincial forest laws (Quebec).

Summary

Provincial governments have been a key driver in the development, implementation, and enforcement of certification in Canada. They have responded

Figure 4.6

The temporal dynamics of certification/policy cycle alignment

Certification policy alignment	Stage of policy cycle	
	Early	*Late*
High	***Ontario*** Directly engaged in standards development and implementation and mandated certification.	***New Brunswick*** Took indirect approach to standards development and implementation but directly mandated certification.
Lower	***British Columbia*** Directly engaged in standards development but took an initial indirect approach to implementation and enforcement prior to change in legislation.	***Quebec*** Took initial hands-off approach to certification development, implementation, and enforcement prior to the 2001 Forest Act and Coulombe Commission.

to certification with varying and increasingly direct co-regulatory approaches, ranging from initially observing certification to cooperating, endorsing, enabling, and even mandating certification. Government rationale for engaging in certification has been similar across regions as a result of similar forest regime conditions. All provinces have participated in certification fundamentally to protect and promote their local forest economies, promote SFM improvements, and maintain market and societal trust in provincial government forest management. The provinces have varied in their certification co-regulatory role, however, in response to three significant factors: industry expectations of the government's role, ENGO advocacy pressure, and alignment between certification and forest policy depending on the stage of the provincial forest policy cycle.

This chapter has shown that industry expectation of government influenced a government's certification response, particularly if the industry presented a strong, united front and the government was sufficiently aware of certification. ENGO advocacy pressure influenced a government's role in certification depending on the issue and on whether the campaigns were local or global in origin. Global campaigns caused governments to assume a more active role at the certification standard development stage in order to protect the provincial forest policy agenda. Advocacy campaigns that focused on issues with a long history in the province (such as spruce budworm in New Brunswick) triggered historic business/government alliances and prompted government support of industry certification efforts. Finally, governments at an early stage of their policy cycle (agenda setting and policy formulation) directly engaged in standards development in order to ensure certification/policy alignment. Confidence in such alignment was a critical factor in influencing governments to directly mandate certification.

Overall, this chapter has shown how forest certification in Canada has gone beyond categorization as a non-state market-driven governance mechanism. Provincial governments have directly co-regulated certification, and government authority has played a critical role in establishing Canada's leadership in forest certification. In the next chapter, we shall see the implications of government enrollment in certification for the governance of state forests across the United States.

5

United States: Enhanced Governance of Certified State Forests

State governments in the United States have been playing an increasing leadership role in forest certification by directly certifying their state-owned public forestland. This has occurred to such an extent that although state forests comprise only 8 percent of US forestlands, they account for over half of the forest area certified to the Forest Stewardship Council (FSC) standard and a quarter of the forest area certified to the Sustainable Forestry Initiative (SFI) standard across the US. It is puzzling that in a country where most forestland is privately owned, state governments are leading the way in adoption of a private forest governance system on publicly owned and managed forestland. Why has there been such an enthusiastic response towards certification on the part of state governments, and what implications has it had for the governance of state forests?

This chapter evaluates the co-regulatory response of state governments to certification, focusing specifically on those states that have certified their state-owned forestland. The findings draw on fifty interviews that I conducted with certification experts and practitioners from government, the forest industry, family forest owner associations, forest auditing bodies, academia, and nongovernmental organizations (NGOs) across the US in 2006-07 (See Appendix A). The chapter presents two main points. First, state governments certified their forests as a result of both market and non-market drivers, including the availability of private foundation funding, environmental NGO (ENGO) advocacy and customer pressure, and the opportunity to demonstrate state leadership. Second, by adopting certification, state governments improved the governance of their forests through greater efficiencies, transparency, and effectiveness in state forest practices and administration.

The chapter begins with a brief overview of the US forest economy and the patchwork of multilevel forest governance and tenure arrangements across the country. After a summary of certification development and adoption in the US, the analysis turns to the role of state governments in certification. In exploring the varied terrain of state-level forest governance, I show

how across different subnational state forest regimes (different tenure arrangements and varied regulatory/voluntary approaches), state governments have increasingly adopted a similar direct co-regulatory response to certification, namely, certifying their state-owned forests. I then evaluate the drivers, implementation debates, and rationale behind state government decisions to directly engage in certification. In particular, the chapter outlines the range of factors influencing state certification response, and the social, economic, and environmental rationale presented to justify the state certification decision. The chapter concludes with an assessment of the governance implications of state forest certification in terms of the implementation challenges and outcomes, and demonstrates that the certification hurdles translated into a range of forest governance benefits. These included important personnel, program, and resource enhancements to the administration and management of state forests.

Variation in US Forestry Regimes

The United States is the world's largest producer and consumer of forest products, and ranks fourth in the world (just behind Canada) in terms of total forest area. The country holds 6 percent of the world's productive forestland and accounts for approximately one-quarter of global industrial wood production and approximately 13 percent of the world's certified forest. The US is also a key player in the development and evolution of forest certification governance. This section provides a brief overview of forestry in the US, the complexity of forest ownership and administration, and the central role of state governments in forest regulation.

US Forests and State Forest Tenure

The US has vast forests that cover a third of its land area (747 million acres), and two-thirds (504 million acres) is classified as timberlands (forest land that is capable of producing in excess of 20 cubic feet per acre per year and is not legally withdrawn from timber production). There is a diversity of forest types distributed across the country, ranging from mixed temperate and high-value second-growth hardwood forests in the East to older coniferous forests on the West Coast, to fast-growing hardwood and softwood forests and pine plantations in the South. While Eastern forests have the highest percentage of timberland comprising their total forest area, most of the wood harvested in the US comes from Western and Southern forests. Over the past two decades, as a result of declining federal harvests in the West, timber production has been shifting to Eastern and Southern forests. Ninety-two percent of the timber harvested in the US is from private forestland.

As shown in Table 5.1, although most of the forested area in the US is privately owned (particularly by small non-industrial private forest [NIPF] owners),[1] there are also sizable public land holdings. For example, the federal

Table 5.1

US forest ownership (in millions of acres)

Ownership	Forestland		Timberland	
Public		319		147
State	63		30	
County and municipal	10		8	
Federal	246		109	
National Forest	(60%)		(88%)	
Bureau of Land Management	(18%)		(6%)	
Other	(22%)		(6%)	
Private		430		356
Family forest (non-industrial private forest)	363		290	
Industrial[1]	67		66	
Total		749		503

1 "Industrial" refers to private land owned by companies or individuals operating wood-processing and manufacturing plants. In this book, I also include investment-based landowner organizations such as real estate investment trusts (REITs), timber investment management organizations (TIMOs), and large private landowners managing forests for commercial production in this category.

Source: USDA 2002.

government owns approximately one-third of the US forests and state governments own 8 percent. State governments across the country generally have at least 100,000 acres of state forest (Figure 5.1), but some states own and manage large public forest estates of over 2 million acres. State forest in the West is primarily trust land that was apportioned at statehood and is still relied upon to generate revenues for educational institutions.[2] Much of the state forestland in the Central Great Lakes and Eastern regions is tax-forfeited land that was acquired (and reforested) by the state in the late nineteenth and early twentieth centuries, after the lands had been either burned or cut over. There is a very small amount of either federal or state-owned public forest in the South.

The distribution of forest tenures across the US is characterized by a complex checkerboard pattern of public and private ownership. Overall, most forests in the East and South are privately owned and forests in the West have higher public tenure. Tenure, however, varies not just according to whether the land is public or private but also according to whether the forests are federally, state-, or county-owned, and whether the private land is held by industrial or small non-industrial, family forest owners. Specifically, the Western states have the greatest percentage of federally owned forest. Small family forest owners hold most of the forestland in the South. In the north-central Great Lakes states, public lands are largely state- and county-owned,

Figure 5.1

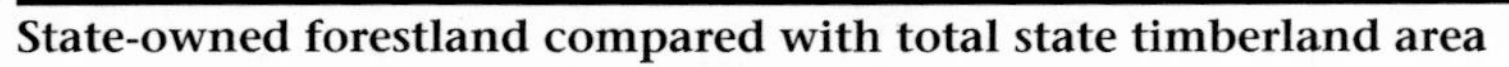

State-owned forestland compared with total state timberland area

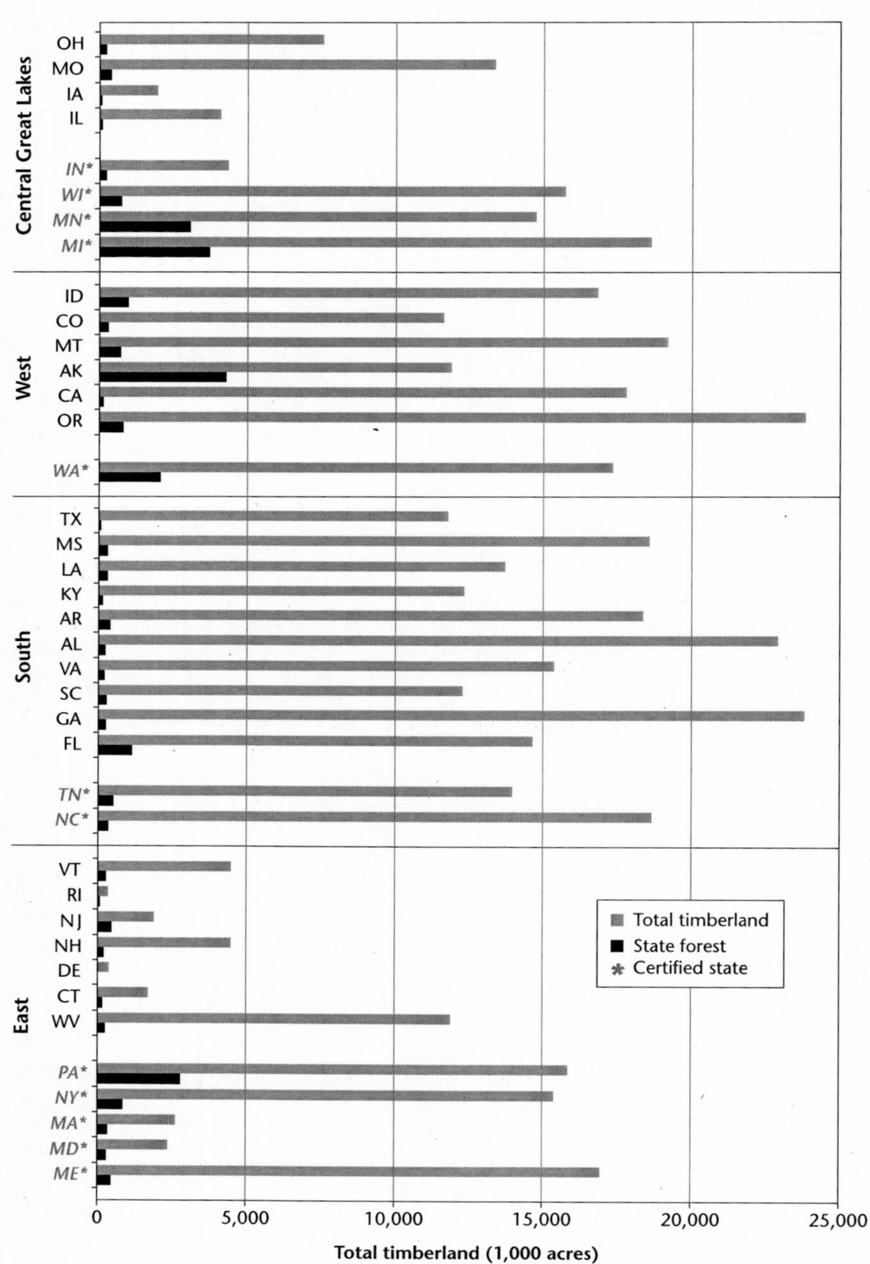

Source: USDA 2002.

and in the northeastern region, there have traditionally been large industrial private timberland holdings (until recent sales). In addition, in these northern regions the state governments own roughly twice as much land as the federal government.

Forest tenure in the US is further complicated by large recent changes in ownership. For example, over the past five years there has been a rapid transfer of ownership of huge tracts of industrial land from large forest companies to timber investment management organizations (TIMOs) and real estate investment trusts (REITs) that, in some cases, have been subdividing and converting the land from forestry to development purposes. Some of this land is finding its way back to public ownership, particularly through trusts and conservation easements.

Understanding this confusing and evolving array of forest ownership in the US is important because tenure is a contributing factor to state forest certification response. Most states with significant forestland holdings have certified their state forests, and changes to industrial land tenure have spurred state governments to certify to set an example for their small family forest owners.

Multilevel Forest Administration

The administration of US private and public forestlands is also far from straightforward. Many different agencies within the federal, state, county, and municipal levels of government have forest management responsibilities, and there is no single law governing forest regulation. The forest management system in the US is multilevel but also generally decentralized, with most forest regulatory authority residing with the fifty individual state governments.[3] Besides managing their state-owned forests, state governments play *the* central role in regulating the management of private forests across the country, and are also the primary vehicle for the delivery of the many forest programs funded by the federal government to assist private forestland owners. Although the federal government owns one-third of US forests, federal agencies have little authority to regulate private or non-federal forestland.[4]

State public and private forests are managed by a lead state forest agency that is typically the Department of Natural Resources (DNR). All states have a state forester who is an executive-level administrator and typically the director of the DNR, although in some states (such as Washington), the state forester reports to a senior deputy or land commissioner. Many states also have independent boards or commissions that play a role in establishing state forest policy.[5] The lead state forest agency is generally responsible for overseeing state parks, wildlife habitat areas, state timberlands, and private forest activities, but states vary in their administrative organization. For example, while all states have separate units for the administration of private

Figure 5.2

Comprehensive state forest practices acts

Alaska Forest Resources and Protection Act
California Z'Berg-Nejedly Forest Practices Act
Connecticut Forest Practices Act
Idaho Forest Practices Act
Maine Timber Harvest Reporting Law
Massachusetts Forest Practices Cutting Act
Montana Notification and Streamside Management Acts
Nevada Forest Practice Act
New Mexico Forest Conservation Act
Oregon Forest Practices Act
Utah Forest Practices Act
Vermont Heavy Cutting and Water Pollution Acts
Virginia Forest Practices Notification Act
Washington Forest Practices Act
West Virginia Logging Sediment Control Act

Source: Ellefson et al. 2007.

forests versus state-owned forests, the delineation of forestry, parks, and wildlife agencies varies.[6]

The role of the state forest agency with respect to private forest activities generally includes administering rules for forest practices covering such areas as reforestation, logging methods, road construction, and other activities that may affect forest management, as well as water quality, wildlife, and other community forest values. With regard to public land, the DNR is responsible for ensuring the sustainable management of state-owned forests, including planning state harvest and timber sales, and managing recreation programs, including state campgrounds and hiking trails. In about half of the states, the state forest agency operates under multiple-use mandates to optimize recreation, timber, and ecological values.[7] The emphasis varies, however. For example, most state-owned forestland in the Western and Midwestern states is *trust land*. It is managed not only to meet state forest regulations but also to earn revenue through timber sales to help fund educational institutions.

Beyond the administrative and organizational differences, US forest governance is further complicated by the fact that state governments differ in their regulatory approaches, employing a different mix of educational, technical assistance, voluntary guidelines, financial incentives, and regulatory measures to promote the sustainable management of forests among private landowners. For example, some states have enacted strongly interventionist forest practices acts to regulate and oversee private forestland

Figure 5.3

The range of US forest policy regimes

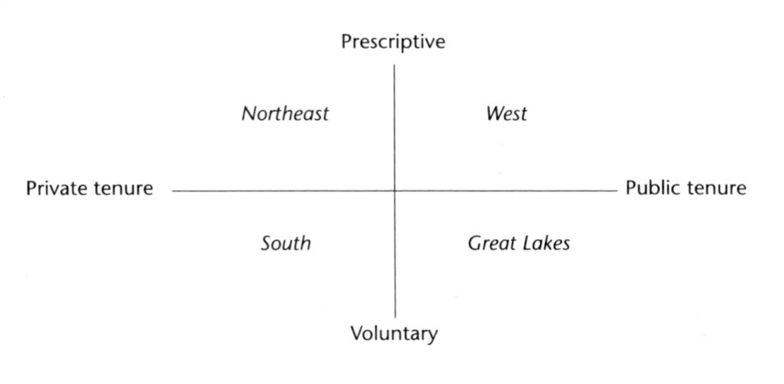

practices (Figure 5.2). Other states rely on voluntary best management practices to guide forest practices, while still other states have taken a minimal interventionist policy approach, essentially leaving private forestland management decisions to the property owners. Overall, the West and Northeast are generally prescriptive whereas the Central Great Lakes region and the South rely on voluntary guidelines. Typically, all states offer extension programs (many funded in cooperation with the federal government and sometimes in partnership with industry) to help family forest owners with education and planning for the sustainable management of their private woodlots.

US Forestry Regimes

Drawing together the institutional details regarding tenure and regulatory approach just presented, it can be seen that there is fundamental variation in the forest policy regimes across the US. As shown in Figure 5.3, this is most evident in comparing the Western regions to the South.[8] The West has higher public tenure (mostly federal land) and strictly enforced forest practice acts, while the South is mostly private (family-owned) land, and state forest policies consist of voluntary best management practices (BMPs) (although these are backed by general state water quality regulations that are mandatory and that can be used for enforcement of specific BMP requirements). The Northeastern region is distinguished by high private tenure and several states have comprehensive state forest practice acts. The Central Great Lakes states also have high private tenure but have less prescriptive regulatory approaches. The forest regimes in the Great Lakes states are also distinctive due to the high relative percentages of state and county-owned forestland.

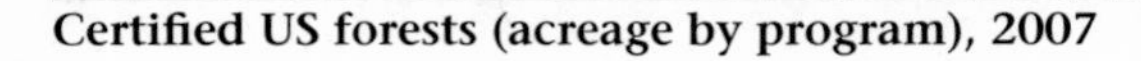

Figure 5.4

Certified US forests (acreage by program), 2007

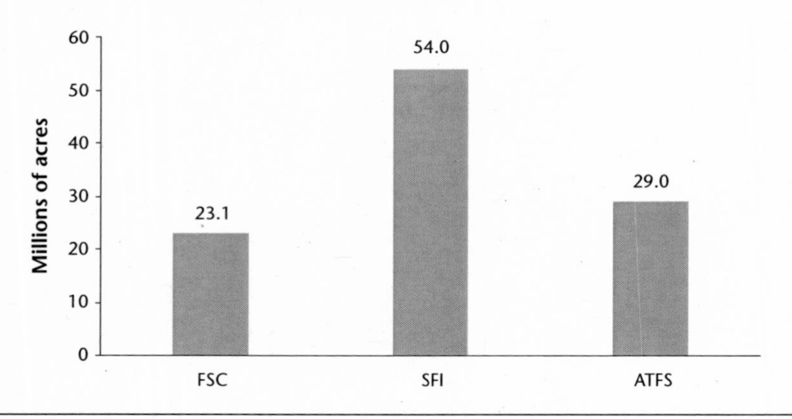

Sources: http://www.fsc.org; http://www.treefarmsystem.org; http://www.sfiprogram.org.

Although forestry in the US is characterized by several distinct forestry regimes, the states face many common forestry challenges, including reduced state budgets, declining forest industry competitiveness and mill closures, the fragmentation of forests and increasing land conversion due to development pressures and the sale of private industrial forestland, and balancing social, economic, and environmental forest values. All states have focused on finding effective means of addressing these issues, and certification has presented an additional governance tool.

Certification Development and Adoption in the US

US Certification Status

As of June 2007, approximately 106 million acres of US forestland had been certified, representing approximately 21 percent of the country's timberland area. While industrial/commercial private forest owners account for most of the US certified forest, state governments lead the nation, with the largest share of FSC-certified forestland. There are three predominant forest certification programs operating in the United States: the FSC, the SFI, and the American Tree Farm System (ATFS) (Figure 5.4).[9] The SFI and FSC programs pertain to all forest owners, whereas the ATFS program is intended for small non-industrial family forest owners. The American Forest and Paper Association (AF&PA) developed the SFI standard, and the FSC US organization has overseen the nine regional FSC standard-setting processes across the country.[10] Industrial forest owners account for most of the SFI-certified forest (73 percent), while state governments are the largest FSC certification

Figure 5.5

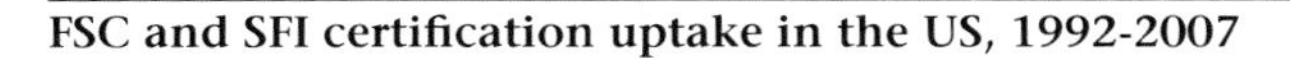

FSC and SFI certification uptake in the US, 1992-2007

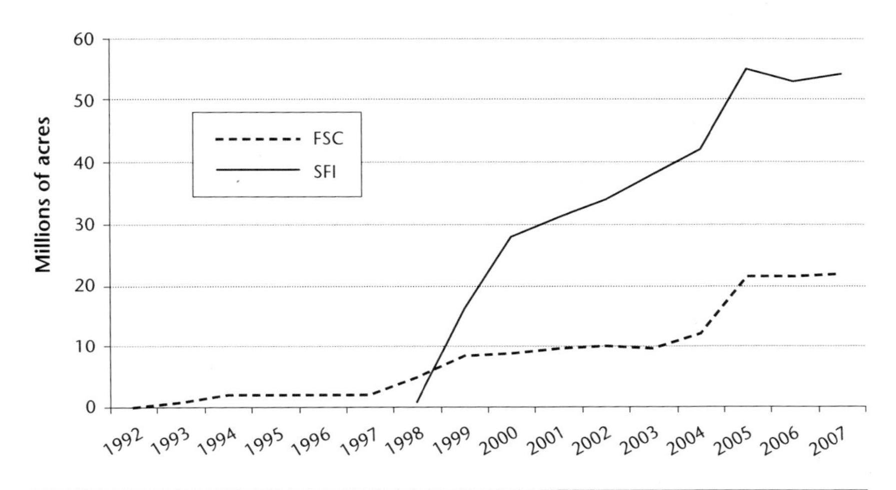

Sources: Alvarez 2007; http://www.certifiedwood.org.

holders (52 percent). Less than 1 percent of non-industrial family forestland has been certified.

Adoption and Evolution of Certification in the US

Third-party certification of US forests has occurred gradually since 1993. ATFS certification has consisted for the most part of encouraging the approximately 53,000 family forest owners already enrolled in the long-standing ATFS program (established in 1941) to seek independent certification under the revised ATFS standard (2000). SFI uptake was predominantly influenced by the AF&PA membership requirement, and FSC uptake has increased slowly as the regional standards developed and gained legitimacy (Figure 5.5).

Enrollment in the SFI program was initiated in 1996 when the AF&PA made its Sustainable Forestry Initiative Principles and Implementation Guidelines mandatory for all its members. In 1998, the AF&PA expanded the guidelines into a standard with verification procedures and third-party certification. At the same time, the industry association initiated the SFI licensing program that allowed non-AF&PA members to enroll in the SFI program. Many of the large industrial forest companies (whether AF&PA members or not), therefore, certified their forestlands in the period 1998-2001. St. Louis County, Minnesota, was the first public agency to enroll in the SFI (1998), and Massachusetts was the first state to enroll (not third-party-certify) its state-owned forests (1999). From 2001 to 2004, there was a gradual increase in SFI-certified forest, with a jump in 2004-05 due to several large state and county forest public land certifications (see Figure 5.5).

FSC adoption in the US lagged behind the SFI in terms of total acreage but has been slowly increasing since 1993, with a considerable spike occurring in 2004-05 largely due to the certification of state- and county-owned forests. The Rainforest Alliance's SmartWood Program (established in 1989) and the Scientific Certification Systems (SCS) Forest Conservation Program (established in 1991) awarded the first FSC certificates in the US in 1993 and 1994, respectively.[11] In 1996, a US contact person was designated by the FSC as the first stage in creating a US FSC national initiative. FSC regional working groups were also established at this time.[12] In this same year, SmartWood issued the first FSC *public land* forest management certification for the Quabbin Reservoir (58,000 acres) in Massachusetts. Between 1998 and 2004, FSC certifications in the US increased only marginally, but during 2004-05, the total acreage of FSC-certified land doubled with the addition of approximately 9.6 million acres of state and county public forestland certifications.

Over the past decade, the various certification systems in the US have continued to evolve. For example, the SFI and FSC programs introduced options to encourage certification among small forest owners.[13] In 2005, the Programme for the Endorsement of Forest Certification (PEFC) international certification program recognized the SFI system. In order to obtain similar international PEFC recognition, the ATFS revised its standards to include mandatory performance measures, written management plans for new and continuing membership, and formal audit training for ATFS inspectors. The ATFS received PEFC approval in August 2008. The ATFS and SFI programs mutually recognized each other in July 2000. In practice, many US customers have inclusive procurement policies that accept SFI, ATFS, or FSC standards, and there is also a trend towards dual-certifying forestland using one combined audit team. State governments have been leaders in fostering the trend towards dual SFI-FSC certification.

Government Role in Forest Certification

The federal government has been largely an observer in forest certification. No federal lands have been certified to the SFI or FSC standards (with the exception of a small area of Department of Defense forestlands at the Fort Lewis base in Washington state, which became FSC-certified in 2002). At the 2005 UNECE Timber Committee meeting in Geneva, the US Forest Service stated: "The US federal government does not intervene in forest certification. It does not act as a standard setting or accreditation body nor does it favor any one certification scheme" (Koleva 2006). The federal government has, however, taken an interest in minimizing the trade implications of forest certification by encouraging the mutual recognition of the various standards, and has also been engaged in developing procurement and trade policies to favour certified, sustainably produced wood products and to discourage illegal logging. In terms of the federally owned national forests, the US Forest Service had considered the prospect of certification for several years, and in 2007 completed a series of pilot tests to assess the feasibility of third-party certification of federal public forestlands.[14] Although the results of the pilot audits were favourable, the Forest Service did not immediately decide to proceed with certification, announcing instead its intention to undertake further consultation. The government was concerned that certification of the National Forest System would be perceived as reinforcing a focus on timber harvest and would add new procedures to an already process-heavy administrative system. The FSC US organization was also undecided and the ENGO community was divided as to the potential costs and benefits of national forest certification, thus reinforcing the Forest Service's reluctance to participate directly.[15]

In contrast to the federal government's tentative response to certification, many state and some county governments have actively engaged in the process. As shown in Figure 5.6, state government certification responses have ranged from passive observance to facilitating certification by providing information and assistance as requested, to enabling family forest certification through land tax incentives,[16] to endorsing certification through state forestland certification and even enforcing certification through state legislation. Only one state (Michigan) has legislated certification, but all states have been closely observing the market dynamics of forest certification and facilitating certification development and implementation by providing information to forest owners, auditors, and certification bodies as requested. State government representatives have also participated in the state-level SFI implementation committees (SICs), and Marvin Brown, the Oregon state forester, has been a member of the SFI board.

As highlighted in Figure 5.6, the most prominent aspect of state governments' role in certification has been the certification of state forests. We turn now to the question of why state governments have chosen to adopt certification on their public land.

The Certification of State-Owned Forests

Over the past decade, state governments have certified a total of 14 million acres of public forest. Although state-owned forests represent just 8 percent of US forests, by spring 2007 they accounted for approximately 23 percent of the country's certified forest area. As shown in Figure 5.7, the twelve states that have certified their forests span the country and encompass several different forestry regions.

Figure 5.6

The role of state governments in forest certification

Spectrum of state government intervention

Indirect ⟶ Direct

Observe	*Cooperate*	*Enable*	***Endorse***	*Mandate*
Keep an eye on certification market developments	Provide information and assistance as requested	Provide land tax incentives for certified family forestland	**Certify state-owned forests**	Enforce certification
All states	All states	Wisconsin the lead state	**12 states have certified their state forests**	Michigan has legislated certification

Figure 5.7

Certified state forests, 2007

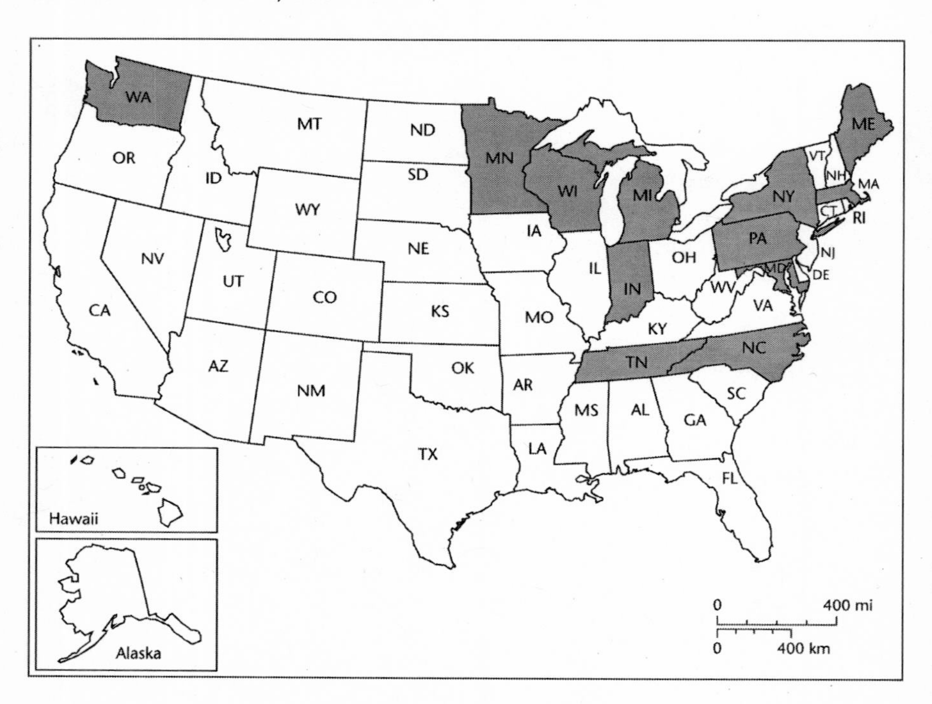

The participating states, in order of certification adoption, are Massachusetts, Minnesota, Pennsylvania, New York, North Carolina, Maine, Tennessee, Maryland, Wisconsin, Washington, Michigan, and Indiana (Table 5.2). Most states pursued dual certification to both the SFI and FSC standards; the exceptions were Tennessee, Massachusetts, and Pennsylvania (FSC only), and Washington state (SFI only).[17]

Timing of State Certification

As shown in Table 5.2, Pennsylvania, Minnesota, and Massachusetts were the early leaders in state certification. Pennsylvania had certified its entire 2.1 million acres of state-owned forest to the FSC standard by October 1998. Before this, the only other public land certifications had been the 1996 FSC certification of 59,000 acres of municipal watershed around the Quabbin Reservoir in Massachusetts, and the FSC certification in the fall of 1997 of 563,000 acres of forest land in Aiken County, Minnesota. New York certified 715,000 acres of its state forests to the FSC standard in 2000. North Carolina, Maine, and Tennessee followed with FSC certifications in 2001-02. All of the early state forest certifications were to the FSC, as the SFI standard did not yet have an independent third-party certification option. In early 2002, Maine became the first state to achieve dual certification to both the FSC and SFI standards. North Carolina achieved dual certification later that year.

In the late 1990s, several states committed to pursuing certification of their state forestlands but implementation was delayed. As a result, between 2003 and 2006 certified state forest acreage jumped significantly, with Maryland, Wisconsin, Washington, and Michigan all achieving certification during this period. Also during this period, Massachusetts and Minnesota extended their initial certified acreage to include a greater percentage of their state-owned forestland. All of these states pursued dual certification, except for Massachusetts, which certified its state forests to the FSC standard only, and Washington, which certified to the SFI standard only. It is noteworthy that although Washington, Michigan, and Minnesota lagged in achieving certification, all three had been working towards certification for over a decade, and all were certifying vast acreages of state forest. By 2006, these states alone accounted for close to 11 million acres, or 78.5 percent of the certified state forest area across the country.

Since the initial wave of certifications in 2000-06, the state forest certification trend has continued. Indiana achieved dual certification of its state forests early in 2007, and in December 2007, the governor of Ohio announced the state's intention to pursue dual certification. Several other states are considering certifying their state lands, including Oregon, which conducted an FSC pre-assessment review of state lands and has reviewed the possibility

Table 5.2

Certified state-owned forestlands, 1996-2007

State	Date	Certification	Area (acres)	Certifier	Notes/status
Massachusetts, Quabbin Reservoir	1996	FSC	59,000	SCS	Municipal watershed.
Minnesota, Aiken County	Fall 1997	FSC	563,000	SmartWood	340,000 acres state forest and 223,000 acres county forestland.
Pennsylvania	Nov 1997	FSC	2.1 million	SCS	In October 1998, 0.9 million acres were added to the initial 1.2 million acres.
New York	Jan 2000	FSC (+ SFI)	715,000	SmartWood	Currently bidding for dual certification (FSC lapsed June 2005).
North Carolina	Nov 2001 Sep 2002	SFI FSC	32,000	SmartWood Plum Line	Withdrew from certification as of April 2007.
Maine	Feb-Mar, 2002	FSC SFI	485,000	SCS Interforest	29,587 acres of Baxter State Park FSC-certified by SCS in 2006.
Tennessee	Oct 2002	FSC	158,000	SmartWood	FSC recertification in 2007.
Maryland, Chesapeake Forest Project	Aug 2003 Apr 2004	SFI FSC	58,000	NSF-ISR SCS	In June 2005, 28,603 acres of DNR forest were added to the original 29,995 acres.
Massachusetts	Apr 2004	FSC	500,000	SCS	Builds on 1998 Quabbin Reservoir FSC certification.
Wisconsin State Forests	May 2004	FSC SFI	490,000	SCS NSF-ISR	439,000 of DNR country land also certified in March 2005.
Washington	May 2005 May 2008	SFI FSC	2.1 million 145,000	BVQI	In May 2005, 1.4 million acres of western state trust lands were certified. In September 2006, 700,000 acres of eastern trust lands were certified.
Michigan	Sep 2005 Dec 2005	SFI FSC	3.75 million	NSF-ISR SCS	
Minnesota	Dec 2005	FSC, SFI	4.8 million	NSF-ISR, SCS	
Indiana	Dec 2006 2007	SFI FSC	150,000	NSF-ISR, SCS	SFI-certified in December 2006 and FSC-certified in July 2007.

of a statewide PEFC certification of all of its public and private forestlands managed in compliance with the Oregon Forest Practices Act. The Southern states are also keeping a watchful eye, with the majority not taking direct action but maintaining a good position to respond if and when big changes occur in the global market.[18]

State Certification Drivers

Among the states that have certified their publicly owned forests, some were early adopters, a few faced challenges that delayed their certification efforts, and others have only recently committed to certification. Factors that influenced state governments to seek and achieve certification when they did were both issue-based and opportunity-based. Issue-based drivers included:

- buyer pressure for certified fibre
- ENGO advocacy and low public trust
- ailing state economy and forest sector
- interstate competition.

Opportunity-based drivers included:

- private foundation grants facilitated by the Pinchot Institute for Conservation (see below)
- market access and potential price premium
- state government leadership.

In addition, there were a few unique cases in which additional drivers such as county and municipal leadership (Minnesota, Massachusetts), university partnership (North Carolina), and land transfer conditions (Maryland) encouraged certification.

It is important to note from this list that many of the factors influencing the states' decisions to proceed with certification such as government leadership, private foundation grants, and ENGO advocacy were not market-based drivers (Table 5.3). In other words, certification of the state forests was not entirely a market-driven initiative.

Pinchot Institute for Conservation Public Land Certification Pilot Audits

The significant early impetus for state forest certification was private foundation funding from several sources (Table 5.4). The Pinchot Institute for Conservation, based in Washington, DC, coordinated the provision of foundation support to state governments in the late 1990s to 2001 as part of its public land certification pilot projects.

In 1996, there were a number of private land certifications in the US but no public land had been certified. The forest sector was unclear how

Table 5.3

Certification drivers by state

State	Issue-based drivers				Opportunity-based drivers		
	Buyer pressure	State economy	Interstate competition	ENGO advocacy	State government leadership	Pinchot Pilot Funding	Market access and premium
Pennsylvania					✓	✓	✓
New York						✓	✓
Minnesota	✓	✓	✓		✓	✓	
Massachusetts					✓		
North Carolina					✓	✓	
Maine	✓	✓		✓	✓	✓	✓
Tennessee				✓		✓	✓
Wisconsin	✓	✓	✓		✓	✓	✓
Washington			✓	✓			
Michigan	✓	✓	✓	✓			
Indiana		✓	✓	✓			✓

Table 5.4

Private funding for certification of state forestland

State	Certification activity	Private funding sources
Pennsylvania	1997-98 FSC certification of all state forest land	Heinz Endowments
Tennessee	October 2001 FSC and SFI assessment audits and October 2002 FSC certification	Pinchot Institute Pilot Project
Maine	June 2001 FSC and SFI audits and February 2002 dual certification of state forestlands	Pinchot Institute Pilot Project
North Carolina	2001-02 SFI and FSC audits and certification of Bladen Lakes State Forest	Pinchot Institute Pilot Project Doris Duke Foundation
Minnesota	1997 Aiken County FSC certification	Rockefeller Brothers Fund
Michigan[1]	1998 forest management plan revisions	Great Lakes Protection Fund
New York	1998 FSC audit and January 2000 certification of state forests	Great Lakes Protection Fund
Washington[2]	September 2000 and 2003 FSC audit and phase 1 SFI audit of western trust lands	Private foundations and the Lanoga Corporation

1 The funding was given to help revise Michigan's state forest management plan to meet the Canadian CAN/CSA-Z809 certification requirements.
2 The state participated in the Pinchot Institute for Conservation's dual certification pilot audit in 2000 but did not move forward with certification at this time and eventually funded its state forest certification out of state revenues.

certification would work for public forest management. The Pinchot Institute launched a long-term study to examine the applicability of independent, third-party certification to forest management on public lands. Seven states participated and several of them went on to achieve certification of their forests after the pilot project.[19] As explained by the North Carolina Division of Forest Resources, "the state had been tracking certification for some time and then Pinchot funding became available to enable the state to proceed."[20] The Pennsylvania Department of Natural Resources similarly emphasized that "without foundation funding and the Pinchot Pilot project, our state certification would have been delayed as it would have been hard to justify the expense at the time given the tight state budgets."[21] By 2001, following the "dot-com collapse" and the overall post-9/11 market decline, Pinchot's private funding sources dried up and several states that had intended to participate either put their certification plans on hold (for example, California) or redirected their efforts towards securing other sources of funding (for example, Washington).

Buyer Pressure and Market Opportunity

Pressure from large forest products customers and the potential opportunity to achieve market advantages for certified forest products were key drivers of state certification. The shifting market dynamics were particularly influential in the Central Great Lakes states, where the buyer threats and market opportunities were very real.[22] The most direct pressure was from Time Inc. (the world's largest buyer of coated paper). The state of Maine had already responded to buyer pressure by announcing its statewide certification initiative.[23] Wisconsin and Minnesota were also among Time's leading paper suppliers and Time communicated to these states that it was prepared to shift its paper sourcing from the region to other locations in order to secure certified product. Time had also demonstrated that it would reward certified suppliers such as the state of Maine for their certification efforts.[24] As explained by a government official from the Minnesota Department of Natural Resources, Division of Forestry, "David Refkin of Time Inc. was telling states to get certified or the company would end up taking their business elsewhere and Minnesota had two mills providing paper to Time. We needed to respond."[25] The situation in Wisconsin was similar: "The paper sector was a major driver of state certification and the Governor jumped on board. He didn't want to lose the 4000 jobs at the local paper mill. Other paper sector jobs at other mills were also at stake."[26]

European customers were also influential in driving state certification, particularly in the significant hardwood-producing states of Pennsylvania and Tennessee. As hardwood customers in Europe had established preferences for certified wood (particularly from FSC-certified forests), these states were interested in the market opportunity to adopt certification on their public lands in order to reinforce the high quality and high market value of their cherry and maple and capture any potential price premium resulting from certification.

Interstate Competition

Keeping certified fibre within the state and keeping pace with other competing jurisdictions were both drivers of state certification. Great Lakes states, Northeastern states, and Southern states all compete to some extent in terms of type of forest products manufactured, similar industry structures, and similar timber species; states that had certified or that were in the process of certifying were therefore particularly influential in driving their neighbours to certify also. This was particularly evident in the case of the Great Lakes states of Michigan, Wisconsin, and Minnesota, which wanted to keep pace with each other. As well, states certified to ensure that state fibre would continue to be processed within the state, and also to meet any growing demand for certified fibre in-state rather than through imports. For example, at the time of Michigan's certification decision, the Department of Natural Resources expressed concerns that the state's central and western Upper Peninsula fibre was going to Minnesota and Wisconsin mills rather than Michigan mills.[27] In Washington, customer demand for certified product led to a growing shortage of certified fibre in the state. To meet market demand, Washington-based manufacturers were starting to look to other states for certified fibre. Washington state forestland certification was therefore part of an effort to ensure an in-state supply of certified fibre.

Ailing State Economies

Ailing state economies and slumping forest sectors caused several states to certify in order to maintain competitiveness and ensure continued market access. Despite program deficits, the states felt that they could not afford *not* to certify given the dire prospect of mill closures and further job losses. As the Michigan DNR commented, "We knew we wouldn't necessarily gain markets by certifying but we'd also lose if we didn't certify."[28]

Beginning in the late 1990s, many states began facing mounting budget deficits. For example, as reported in the opening paragraph of the 2002 *Fiscal Survey of States,* "nearly every state is in fiscal crisis. Amid a slowing national economy, state revenues have shrunk at the same time that spending pressures are mounting ... creating massive budget shortfalls."[29] Most states were forced to make program cuts, including layoffs in their forestry agencies.

The forest sector also experienced an economic downturn during this period. Mill closures and job losses were common across the US. Concerns about the depressed national economy and uncertainties about the viability of state forest sectors were particularly pronounced in Washington, Maine, and the Great Lakes states, and played a key role in their decisions to proceed with state forest certification. For example, Michigan had been hard hit by the slump in the Midwestern manufacturing sector in the 1990s and had suffered huge losses in the forest industry.[30] As a result, the Healthy Forest Bills were introduced in 2004 to revitalize the Michigan forest products industry, specifically by accessing the large volumes of maturing state timber and by ensuring market access through certification. In particular, House Bill 5554 (Sustainable Forestry on State Forestlands) amended Part 525 of the Natural Resources and Environmental Protection Act, requiring that "by January 1, 2006, the DNR shall seek and maintain forestry certification by at least one credible, non-profit, non-governmental certification program." The legislative mandate put the department on the fast track to achieve dual certification by December 2005.

Minnesota was also hit by a major economic and forest sector downturn during 2000-03. Employment dropped to just under 42,000 jobs from 54,000. Harvests decreased, stumpage prices increased, and fibre imports increased. In 2003, a Governor's Task Force on the Competitiveness of Minnesota's Forest-based Industries was established, and one of its key recommendations was that the DNR certify all state-owned timberlands by 2005 (Minnesota DNR 2003).[31] In Wisconsin (the largest paper-producing state), a key driver in the decision to proceed with state forest certification was the governor's particular concern about the potential loss of any of the 35,000 primary forest products paper sector jobs in the state, and the prospect that certification would stimulate the economy. In order to fast-track an increase in certified state fibre, Wisconsin picked the "lowest hanging fruit" – the 500,000 acres of state-owned forestland. The Wisconsin Department of Natural Resources had already been implementing forest management criteria and indicators so it knew that the state was prepared to meet certification requirements, in contrast to the uncertainty of county land or private non-industrial land certification. Maine faced similar economic pressures. In July 2003, following the loss of approximately 31 percent of its forest products jobs in the 1990s, the governor launched a statewide forest certification initiative to stimulate and "regrow" Maine's forest sector by distinguishing its products in the marketplace while improving forest management on the ground (Maine 2002). In Washington, increased log costs and declining timber prices were putting increasing pressure on state forest management programs. Although not the key driver of state certification, the need to maintain state competitiveness and ensure continued market access for state timber sales through certification contributed to Washington's active engagement in the FSC and SFI certification processes beginning in 1997, and to the state's SFI certification in May 2005 and FSC certification in 2008.

Public and ENGO Advocacy Pressure

Increasing public concerns about non-timber forest values on state forestland and ENGO advocacy were drivers of state certification in all states, particularly in Maine, Tennessee, Washington, Indiana, and Michigan. Most state-owned forestland is either trust land with a mandate for revenue generation or tax-forfeited land that was cutover and abandoned at the turn of the century and subsequently replanted by the state for future harvest. ENGO advocacy pressure in the US has therefore focused for the most part on national forests, not state forests. As cities have expanded, however, and as values shift towards preserving and enhancing recreational and ecological forest benefits, citizens and environmental groups have been calling on state governments to demonstrate commitment to sustainable forest management on publicly owned forestland.

ENGO pressures regarding certification generally fell along a spectrum of interests. Some groups, such as ForestEthics and the WWF, pursued market campaigns to encourage FSC-only certification as a means of rewarding leadership and continual improvement in sustainable forest management practices and to build capacity in the FSC program. Other groups, such as the Sierra Club and the Dogwood Alliance, actively tried to prevent the certification of state forestland (including FSC certification) as they wanted to discourage a potential "licence to harvest" on public land. This heightened ENGO focus on state forest practices ultimately served to drive certification, however. For example, in Tennessee, in response to strong pressure from the Natural Resources Defense Council (NRDC) and local ENGOs, particularly over the Cumberland Plateau forest, the state government accepted the Pinchot Institute's offer to participate in the public land certification pilot project. The NRDC supported the FSC, so Tennessee went with FSC certification. The Tennessee state forester described FSC third-party assessment of state land as the only way to respond to the ENGO pressure.[32]

Citizen concerns and ongoing ENGO campaigns in Washington acted as both a deterrent and a driver. Environmental groups actively encouraged the state to seek FSC certification, and FSC audits of the state trust lands were conducted in 2001 and 2003. Among the audit conditions was a requirement that the Department of Natural Resources recalculate its allowable annual cut and reduce harvest levels by extending rotation ages or increasing green-tree retention. The state was already in the midst of its sustainable harvest recalculation process, however, and FSC certification was deferred. In September 2004, when the state Board of Natural Resources adopted the Sustainable Harvest Calculation and the commissioner of public lands announced a plan to increase harvest on western trust land by 30 percent, environmental groups sued the state for not adequately considering the environmental consequences of the increased harvest.[33] The court challenge further delayed FSC certification. Washington achieved SFI certification in 2005. Environmental groups have continued to encourage the state to FSC-certify all 2 million acres of its state-managed forests, and in March 2007 the land commissioner announced that the state intended to seek FSC certification for 141,000 acres of its western trust lands. Overall, ENGO advocacy has been an ongoing driver of certification in Washington. In such a politically charged environment, the government recognized both the need and the opportunity to demonstrate the state's sustainable forestry practices through some form of independent verification.

In Maine, the governor's July 2003 commitment to certification was prompted by a series of citizen referendums in the 1990s that were critical of state forest practices and that called for a ban on clearcutting. State certification was viewed as way to stem the tide of public distrust by demonstrating the state's sustainable forest practices through third-party verification against an international standard. In Indiana, concern over a plan to increase state forest harvest levels caused environmental groups to be very vocal with the FSC auditors prior to approval of certification for the state forests. This encouraged the state to pass its FSC audits and demonstrate its ongoing commitment to forest stewardship. Finally, in Michigan, environmental groups lobbied effectively for certification to be legislatively mandated. Thus, Michigan became the only state to adopt a legislative requirement for ongoing certification of its state-owned forestland.

State Government Leadership

Besides external factors such as Pinchot Institute funding, buyer pressure, ENGO advocacy, adverse economic conditions, and/or the influence of competing jurisdictions, impetus for certification, especially with respect to timing, came from strong individual leadership from within state governments themselves. This was particularly true in the case of state foresters in Massachusetts (Bob O'Connor), Minnesota (Gerry Rose), Wisconsin (Paul DeLong), North Carolina (Stan Adams), and Pennsylvania (Jim Grace). Supported by committed staff, these individuals persistently championed certification from within their respective departments. Both Wisconsin governor Jim Doyle and Maine governor John Baldacci also decided early on to move forward the overall certification of public and private lands in their states. In North Carolina, the Department of Forest Resources encouraged state certification leadership through an innovative cooperative partnership with the Department of Environment and Natural Resources, Duke University, and North Carolina State University. In all these states, the efforts and support of a few key visionaries in government were key to making certification happen when it did.

In several states, the government's leadership was intended not just to satisfy buyer demands and gain market access but also to address the issues of private land conversion and fragmentation and encourage family forest owners to certify. The state would have had difficulty getting family forest owners to certify if certification had not been adopted first on publicly owned forestlands.

Industry Expectations of the Role of State Governments

Industrial forest producers across the US had differing views on whether state governments *should* be engaging directly in certification, and these expectations had a variable influence on state certification.[34] Some forest companies asserted that state certification did not matter, others were vehemently opposed to it, and others strongly supported it (Table 5.5).

Table 5.5

Industry perspectives on state forest certification

Position on certification of state forests	Rationale
State certification is necessary.	State forests are an important supply of certified fibre to meet customer demands.
State certification does not matter.	The company does not rely on the purchase of state fibre; mills are facing little market pressure for certified product; and the acreage of state forest-land holdings versus private ownership is very small in the region.
State certification benefits state forest policy administration.	State certification enhances the delivery, accountability, communication, and continual improvement of state forest policy.
State certification should not be adopted.	Governments have their own processes and should not be adopting someone else's.

For example, Southern producers did not think that state certification mattered as there was little state-owned forestland in the South and little market pressure to certify. Companies that relied on sourcing of public fibre from state forests (particularly in the Great Lakes region) supported certification, explaining that it increased the availability of certified supply and also helped to improve the delivery of state forest programs.[35] Companies with significant private land holdings were either unsupportive (since additional certified fibre represented competition) or thought that state certification did not really matter one way or another. Indifference to state certification was particularly true of companies operating in the West that did not rely on state fibre because of export restrictions placed on public timber by western state governments. Companies that were very dependent on purchases of private non-industrial fibre across US forest regions generally supported certification of state forests but were more concerned with the supply-side issue of encouraging greater family forest certification. Finally, companies with operations in highly regulated states (such as Maine, Washington, and Oregon) commented that state governments should be "staying out of certification" – that "governments have their own process and shouldn't be adopting someone else's." In particular, the concern was that government engagement in certification might lead to mandatory certification of private land. Overall, the impact of forest company expectations on governments' certification response was uneven across the certified states due to the different tenure and regulatory arrangements and variability in company fibre

Figure 5.8

Evolution of state certification drivers

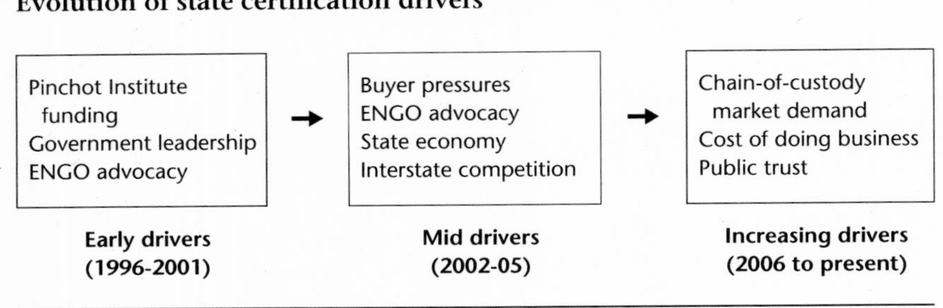

sourcing, and therefore did not have as great an influence as that of more consistent and prevalent factors.

The Evolution of State Forest Certification Drivers

The drivers of state certification have not been static. Over the past fifteen years, they have evolved as certification systems have matured and markets have slowly developed (Figure 5.8). For example, for the states that certified their forests prior to 2001, the key drivers included ENGO advocacy, the availability of Pinchot Institute funding, state government leadership, and the potential for distinguishing state forest products in the marketplace. As certification gained in acceptance and adoption, direct buyer pressure became an increasingly significant driver. And as state certifications increased, interstate competition also emerged as an important influence on state certification.

At present, demand-side buyer pressures are becoming a prominent driver as chain-of-custody certifications increase and buyers increasingly adopt sustainable forestry procurement policies that favour certified forest products. Market benefits through sufficient price premiums have generally not materialized, but certification is gaining market acceptance as it is increasingly viewed as a necessary cost of doing business. As forest certification gains legitimacy through increased uptake, it is becoming an important means of achieving greater accountability and, for governments, of demonstrating responsible forest management and building public trust.

State Certification Implementation Debates

The timing of certification of state forests was influenced not only by the prevalence of the various drivers but also by the debate about whether and how to actually proceed with such certification. The three most commonly debated issues were certification financing, selection of the certification program, and policy sovereignty. In other words, could the state afford certification? Which certification standard should be pursued? And would certification threaten the policy authority of the state?

Financing Certification

A fundamental consideration for many states in deciding whether to pursue certification was simply whether they could afford it. As discussed above, many states were coping with increasing budget deficits and declining forest sector economies. Many government departments questioned the costs of certification and asked how the expense could be justified, given the uncertain and not necessarily measurable benefits. For several states, once the decision was made to proceed, it became a question of resource allocation – whether the state was even in a position to take on certification costs given the large budget deficit and recent large staff cuts. The states were unsure about the actual costs of implementing certification on state forestland, and about whether to consider certification an economic opportunity, a market necessity, or an unjustified (perhaps premature) expense.

Determining the costs. Early certification adopters lacked information about the challenges and benefits of certification as well as the costs of implementation. In order to gain knowledge, several states took advantage of available private foundation funding (particularly through the Pinchot Institute's public land certification pilot projects) to participate in studies to determine the financial impact and learn more about the economic implications of certification. Certification costs were subsequently calculated and compared for the North Carolina, Minnesota, Pennsylvania, and Wisconsin pilots.[36] For example, Dr. Fred Cubbage and his team at North Carolina State University calculated that the total direct costs of the North Carolina Division of Forest Resources 2001 SFI and FSC audit assessments were $0.54 per acre for SFI and $0.72 per acre for FSC (Cubbage et al. 2003).[37] The 1997-98 Pinchot Institute pilot project calculated that the costs of FSC certification assessments for Minnesota and Pennsylvania were $0.09 and $0.10 per acre, respectively, with an additional cost for licensing and ongoing annual auditing of $0.01 to $0.02 per acre (Mater et al. 1999).

States that certified their forests later (2005 onward) not only benefited from the cost information generated in the earlier Pinchot Institute studies but were also able to decrease their costs by better coordinating the audits for the various certification standards. For example, the Indiana state forester explained that his department achieved economies of scale by seeking bids on all three audits (SFI, FSC, and ATFS) together as a package. They then had an efficient system in place for future audits, including a forest carbon auditor.

Justifying the expense. When third-party forest certification emerged in the early 1990s, it had no precedent and its market implications were uncertain. Did it present an economic opportunity or was it a necessity to avoid a market penalty? As summarized in Table 5.6, those in favour of state forest

Table 5.6

Arguments for and against incurring certification expense

Certification represents an ...	Yes	No
Economic opportunity	Can distinguish ("brand") state forest product, create and capture new markets, and obtain price premiums	• No market premiums • Insufficient state-certified supply to create and capture new markets • Others already certified – no longer a unique market position
Market necessity	State mills and large buyers are increasingly demanding certification, so state will lose market access if not certified	• Little demand for certified state fibre
Justified financial expense	At a minimum, state will sustain forest productivity and revenue flows by maintaining market access	• No price premium, so no measurable return (other benefits hard to quantify) • Expense premature as certification markets not yet developed

certification argued that there was economic opportunity in that the state could certify to distinguish its forest practices and forest products in the marketplace ("state branding"), create and capture new markets, and obtain price premiums. Those opposed to incurring any additional costs argued that there was little economic opportunity in certification as market premiums did not exist, there was an insufficient supply of state-certified fibre to create and/or capture new markets, and any window that might have existed for creating a unique market position had closed because so many private industrial landowners had already certified their forests.

Proponents of certification argued that the risks of not certifying were more certain. Evidence was presented that mills within a state, as well as large forest products buyers outside the state, were increasingly demanding certified products and that the state's forests would lose market access if they were not certified. Although this was a convincing argument in many states, in certain regions (particularly the South) there was little evidence of increasing market demand for certified state fibre. The Tennessee Division of Forestry explained: "We're all thinking about certification but not worrying too much. We know that we have so much wood in the South from so many small landowners that state certification is not an immediate concern."[38]

Overall, those opposed to state certification argued that it would be a premature expense as certification markets had not yet developed. The costs could not be justified because there were no price premiums and so there would be no measurable financial returns. In response, proponents explained that, at a minimum, state forest certification would sustain productivity and revenue flows by maintaining market access.

States that went on to certify their forests justified the costs largely on the basis of risk avoidance and expected value. They feared putting the local industry in worse jeopardy, and hoped that new markets and/or a price premium would materialize. It is worth noting that as certification systems mature while significant price premiums have yet to develop, justification of the certification expense continues to present a challenge for some states as they seek their certification renewals. Increasingly, in the absence of clear market price signals, the financial justification for certification is simply that it has become a necessary cost of demonstrating sustainable forestry.

Selecting the Standard

In the early years of certification in the US, the FSC was the only choice if forest owners wanted an independent third-party certification audit; the first certified state forests (in Massachusetts, Minnesota, Pennsylvania, and New York) therefore had FSC certification only. In July 2000, however, when the AF&PA expanded its SFI program to include an independent audit component, debate arose over which standard to pursue and/or maintain.

As outlined in Chapter 3, forest producers and consumers have experienced considerable confusion over the merits of the various standards. The SFI was the US national program, developed and promoted by the domestic forest industry, as opposed to the FSC, which was internationally recognized and accepted by ENGOs. State governments debated how to satisfy both the environmental and industrial constituencies. Various organizations conducted studies to learn more about the distinctions between the programs offered in the US. In 2001, Home Depot, FSC US, and the AF&PA sponsored the most comprehensive study. Conducted by the Washington, DC–based Meridian Institute, it provided a comparative evaluation of the differences and merits of the various programs (Meridian Institute 2001). Several states relied on the comparative information in this study to aid their certification decision, ultimately dual-certifying in order to have a balance of system and performance elements.

In 2001, in response to the availability of SFI third-party verification audits, the Pinchot Institute broadened its public land certification pilot projects to include additional pilot investigations of state forestland *dual certification*. Pinchot offered to cover the costs of the dual-certification audits if the state agency would provide a "reverse assessment" – an evaluation of the audit standards and processes of the SFI and FSC certification programs. Maine,

Figure 5.9

Dual-certification status of US states, 2007

Dual certification		*FSC only*	*SFI only*
Maine		Pennsylvania	
Maryland		Tennessee	
Wisconsin		Massachusetts	
Michigan			
Minnesota			
Indiana			
North Carolina (2001)	→	North Carolina (2006)[1]	
New York (2007)	←	New York (2000)[2]	
Washington (2008)	←	←	Washington (2005)

1 North Carolina was dual-certified (2001-06) but dropped SFI in 2006 following its recertification audit, as the audit found that the state had not adequately demonstrated continual improvement or adequately calculated the state's allowable annual cut, and the state faced budget constraints related to maintaining both standards. In April 2007, the state dropped its FSC certification as well.

2 New York was FSC-certified at first (2000-05), and achieved dual certification in 2007.

Tennessee, North Carolina, and Vermont (federal land), as well as two universities (Duke and North Carolina State), all signed on. Maine and North Carolina went on to dual-certify their state forestlands in 2002.[39] The Pinchot-supported dual-certification pilots established the trend for the majority of states that followed. For example, Maryland, Michigan, Wisconsin, and Indiana all dual-certified their forests, and Minnesota and New York went from FSC certification only to dual SFI and FSC certifications (Figure 5.9).

Beyond the Pinchot Institute impetus, state governments dual-certified because there was a potential for satisfying a greater constituency and because the extra costs and effort were minimal. The Minnesota Department of Natural Resources explained that it went with dual certification to discourage critics of each of the standards. It wanted to build credibility with all stakeholders, and pursuing both standards entailed only a small incremental cost.[40]

Debate over whether dual certification would achieve greater market access resulted in different conclusions in different regions of the country. For example, Pennsylvania and Tennessee pursued the FSC standard only as they had high-value hardwoods going to European markets (which preferred FSC certification). Washington state, however, viewed the SFI program as the most "positive market orientation," as it exported almost no wood to Europe. The Central Great Lakes states argued that dual certification would cover all bases: "After much debate we ended up going with dual-certification because it would address multiple stakeholder interests and preserve, and hopefully even expand domestic and foreign markets."[41]

In subsequent years, the potential market advantages of one standard over another became less convincing with increased convergence of the standards and mutual recognition through inclusive customer procurement policies (see Chapter 3). The SFI gained international recognition through the PEFC program and the FSC US regional standards demonstrated adaptability to local forest conditions. Many of the SFI-certified industrial forest companies commented that the market accepts either standard, so why have both?[42] Consequently, some states, such as North Carolina, dropped their dual certification, whereas for other states, such as Washington, increased convergence lowered hurdles and encouraged dual certification (see Figure 5.9).

Policy Sovereignty

In several states, perhaps the greatest point of debate regarding whether and how to proceed with the certification of state-owned forests was the question of policy sovereignty. Was the government subverting its authority and "handing over the policy reins" by directly endorsing and adopting a set of private rules for the management of its public forests? The debate was particularly pronounced in states with non-discretionary forest regulatory programs, such as in the Northeast and Pacific Northwest forest regions. There were four main areas of concern: policy alignment, control, flexibility, and necessity. Were certification requirements consistent with state policies and programs? What influence would the government have over certification requirements? Would certification systems permit the flexibility required for public forest governance processes and decisions? And was certification even necessary if state forest laws were already comprehensive?

For example, Washington state delayed its FSC certification as the requirements and the timing of the auditor requests were viewed by its Department of Natural Resources to be out of alignment with state forest objectives and processes.[43] The DNR was not looking for a new "policy master"; further, the DNR stressed that it was accountable to the Forest Practices Board (an independent forest agency chaired by the commissioner of public lands) and not the certification bodies. Also, because of the legal land trust relationship, the state could not "pledge allegiance" to an independent certification body. The DNR argued that before proceeding, certification had to be consistent with existing public responsibility, and not a new hurdle or a substitute for political accountability. The Michigan Department of Natural Resources also had fears and concerns regarding sovereignty. It worried that with its statutory requirement to maintain the certification of state-owned forestlands, the government would lose some policy flexibility in terms of being able to accept or reject rules generated by a private or nonprofit entity. Within the department, there were concerns that the legislated state forest certification would raise questions about who was actually leading forest policy in the state, and that the state might be laying itself open to increased court challenges resulting from inconsistent language between the certification standards and state forest policy.

The states also debated the pros and cons of the dynamic nature of the certification standards. On the one hand, there were fears that certification was an "elastic ruler" and that if the state committed to certification, the requirements might unexpectedly ratchet upward, creating possibly undesirable expectations regarding state forests and the state government. On the other hand, the ongoing revisions to the standards could mean that the certification rules and processes would remain adaptive and responsive to changing forest conditions and values. In addition, some private forest owners (particularly in Maine) were concerned that although certification was voluntary, if the government certified state forests it would just be a matter of time before the voluntary rules worked their way into regulations – that it was a potentially slippery slope.

A final significant point of debate, particularly in states with comprehensive forest regulatory programs, concerned why certification was even necessary, given that state forest practices already exceeded certification requirements. For example, in Washington, many felt that the state forest practice rules and Habitat Conservation Plan (HCP) offered strong enough environmental protections so that certification was unnecessary. The governments of California, Idaho, and Oregon wondered whether it would not be more prudent to just "brand" the state forest practices under a unique statewide SFM certification label and conducted gap assessment studies comparing state forest policies and regulations with certification requirements.[44] Oregon rejected the SFI and FSC certification programs largely because it believed that the FSC certification auditors had tried to overly influence the direction of state forest management.[45] Instead, it explored the possibility of PEFC-certifying its comprehensive state forest regulatory program (Pinchot Institute for Conservation 2006).

State governments that proceeded with certification countered the various sovereignty concerns with the fundamental argument that certification was not a substitute or competitive threat to state forest management objectives or authority, but rather a complement to the existing forest regime – that certification was a means to verify, demonstrate, reinforce, and be rewarded for a high level of state forest practices.

Rationale for Certification of State Forests

State government rationale for certifying state-owned forests included a range of economic, environmental, and socio-political reasons (Figure 5.10). The four most common justifications were:

- State certification is a response to global market trends and will improve state forest industry competitiveness.
- State certification will set a leadership example for the state's many private landowners.
- Third-party audit verification will build public trust in state forest practices.
- State certification will demonstrate, reinforce, and/or improve state forest management practices.

Economic justifications focused on maintaining market access and the opportunity to improve the competitiveness of the state forest industry by keeping pace with domestic and global market trends towards certification. The environmental rationale included certifying state forests as a means of demonstrating, reinforcing, and achieving recognition of the state government's forest stewardship commitment to a high level of state forest management and forestry practices. In other cases, the states emphasized that certification would provide an opportunity to achieve better forest management practices through third-party feedback. They also asserted that state forestland certification would enable them to take a leadership role in encouraging private landowners to certify and/or improve their SFM practices. The socio-political rationale for certifying state forests included enabling better public education and ultimately building greater public trust and credibility with citizens and environmental organizations. As is explored next, state governments faced a range of challenges and benefits in working towards the realization of these expectations.

Figure 5.10

Rationale for certification of state forests

Economic: industry competitiveness

- Maintain market access.
- Improve state forestry competitiveness.
- Respond to global trend towards certification.
- Keep pace with or stay ahead of other states.

Environmental: forest stewardship

- Improve state forest management.
- Demonstrate and achieve recognition of existing state forest management practices.
- Provide forest management leadership and set an example for private landowners.

Socio-political: public trust

- Increase citizen understanding, engagement, and support.
- Build credibility with ENGOs.

Figure 5.11

Challenges in implementation of state forest certification
The bars indicate the extent to which the certified states identified each of the challenges. Interviewee responses were based on an open-question format rather than a structured survey. The breakdown of response by state is provided in the summary table in Appendix C.

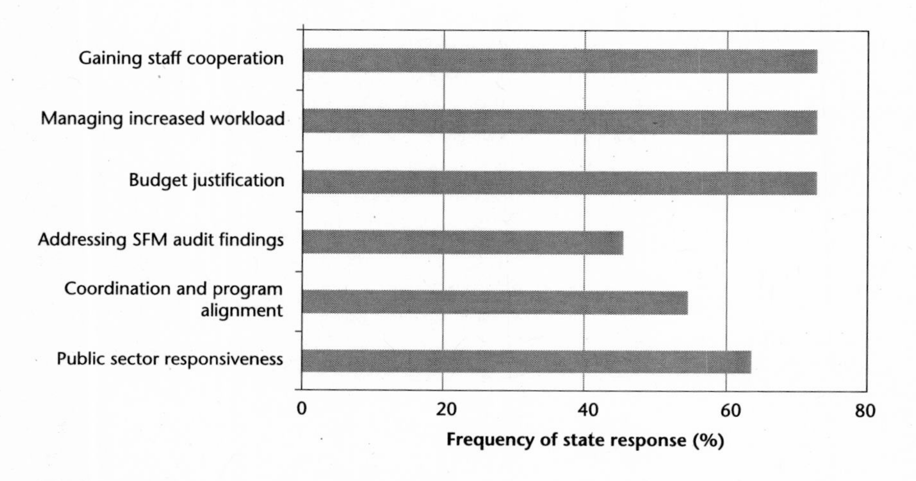

The Governance Implications of State Forest Certification

Implementation Challenges

State forest agencies identified six main hurdles that they had to overcome in implementing certification on state-owned forestlands: gaining staff cooperation, managing an increased workload, justifying the increased budget, addressing SFM audit findings, coordination and program alignment, and public sector responsiveness (Figure 5.11).

Gaining Staff Cooperation

Perhaps the greatest upfront implementation challenge for many state forest agencies was gaining full employee cooperation. Initially, staff were wary of certification and resisted being told what to do by an outside party. Employees did not like the feeling of "someone looking over their shoulder" and did not want certification dictating their jobs. There was also fear that certification was "the camel's nose under the tent" – that certification was a small indicator of much larger changes that were yet to be revealed.

Another challenge was that once they became engaged in the certification process, it took a while for employees to treat certification as a regular departmental activity rather than a different, separate process. Several states faced the particular difficulty of integrating certification into their field forester work programs. In order to facilitate implementation, many state agencies offered certification education programs for their staff. For example, Minnesota implemented an education program to increase staff awareness and help staff realize that certification would help on-the-ground forest practices. As the Minnesota forest certification program coordinator further explained, their wildlife managers became more accepting of certification when they realized through the staff education program that certification included helpful indicators such as those for tracking habitat.

Staff support and morale generally increased after the initial certification audits. Employees realized that the audit findings not only acknowledged their good work but also highlighted gaps that in most cases the department already knew needed fixing. For example, the Michigan Department of Natural Resources explained that the auditors did a good job; when they found a corrective action request, the staff reaction was typically, "Yup, that needed fixing or improving." Other states acknowledged that although initial acceptance was not uniform, staff were now saying that certification was good for them. They knew that they were not doing things as well as they could, and that certification was making things better.

Managing Increased Administrative Workload

A significant implementation challenge for many states was the increased administrative workload. In particular, certification required the formal documentation of current policies and programs, as well as up-to-date training and monitoring records, and continual improvement reporting. The states commented on the high degree of ongoing documentation involved, including upgrading documentation to the level required for audit evidence and developing new policies and procedures to cover inadequately addressed areas. The challenge was not just to develop many new forms and procedures but also to train employees to use the forms, follow the new procedures, and keep up-to-date records. Several states struggled to allocate the additional staff time required, given their tight budgets and reduced manpower. As a result, certification auditors in several instances included as a corrective action request sufficient staffing levels to enable state agencies to achieve SFM programs and objectives. They did not want to see states fail their re-certification audits and lose their certification status due to inadequate human resources.

Budget Justification

Although several states had their initial certification audit costs covered by dedicated state timber revenue, private foundation dollars, and/or additional funds provided through the governor's office, securing the resources needed to address certification audit findings was an ongoing challenge. With already

Table 5.7

Summary of FSC corrective action requests for state forestland

State	Area (thousands of acres)	Auditor	Total corrective action requests	Silviculture action requests
Massachusetts	500	SCS	17	0
Maine	485	SCS	13	3
Pennsylvania	2,100	SCS	12	2
Minnesota	4,840	SCS	14	3
Michigan	3,750	SCS	13	1
Wisconsin	490	SCS	9	0

Source: Seymour 2006.

tight budgets, uncertain market benefits from certification, and the need for additional infrastructure to support certification implementation, the states struggled to come up with convincing financial justifications to get the additional public funds. State purchasing processes emphasized cost/benefit justification and, as explained by the North Carolina Division of Forest Resources, "it was hard to justify an expenditure with no return on investment other than learning."[46]

Addressing SFM Audit Findings

The challenge of addressing audit findings concerning improvements to sustainable forest management practices varied among the states. Some states had many audit findings whereas others had state forest programs that already met most of the certification requirements. There were different corrective action requests for each state. For example, in comparing the FSC certification audit findings for the Northeastern and Great Lakes states, Dr. Robert Seymour at the University of Maine identified a varying number and range of audit findings, particularly with respect to silviculture practices requiring improvements (Table 5.7).

The degree of certification implementation challenge was largely affected by whether the state had conducted pre-assessment audit(s) early on to assess its certification preparedness. The pre-assessments determined whether the state was a good candidate for certification, and identified and enabled the state to address any major gaps before undertaking its full certification audit. Based on the pre-assessment findings, some states decided not to proceed with certification (Tennessee – SFI; California, Oregon, and Washington – FSC) while others delayed certification (Wisconsin, Minnesota).

Meeting certification SFM requirements was not a great challenge for a few states, as they had talked to other states that had already certified and knew what to expect. Some states anticipated the audit findings because they were already action items on the state's forest management priority list (for example, having an inventory broader than timber reviewed on a ten-year basis). For still other states, certification was not viewed as a significant hurdle as the responsible state agency had always managed forests to a broader mandate than simply getting timber to the mill cost-effectively; non-timber values were already being addressed, so there were no real challenges in meeting certification indicators.

The states that had a large number of SFM audit findings grouped the certification implementation challenges with respect to completing required management plans, shortening inventory cycles, and factoring in additional non-timber considerations. For example, changing to a ten-year inventory cycle from a twenty-year cycle demanded process improvements and more staff time. Updating and expanding the management plan to include all state forest divisions and non-timber considerations within the context of a landscape-level framework also required more technical and human resources.

Departmental Coordination and Program Alignment

Agency coordination has been a historical administrative hurdle with respect to the delivery of state forest management policies and programs and was a particular challenge for certification implementation. For example, although responsibility for state forest management typically resides in a lead forest agency, such as the Division of Forestry within the Department of Natural Resources, many non-timber issues such as wildlife management, soil and water quality, biodiversity, and state parks are shared across divisions and even across departments. In these cases, certification demanded a coordinated departmental response to meet the requirements that fell outside the Division of Forestry's core responsibility. For example, the Pennsylvania Department of Conservation and Natural Resources (DCNR) explained that although deer management was a major corrective action request, state wildlife management fell under a separate independent agency, not the DCNR. The Michigan Department of Natural Resources explained that many corrective action requests required the response of the entire department. For example, improving stakeholder involvement cut across all divisions, and in order to address the challenge, the department established a Forest Certification Implementation Team to facilitate coordination among them.

In addition, several states faced certification coordination challenges not just at the organizational level but also at the program level, in terms of ensuring the alignment of certification requirements with various state plans and processes, many of which had different timelines and spatial scales. Most state forestry departments did not have a single planning cycle but rather multiple plans and multiple cycles. Some plans had long time frames and others short, and none took priority. The challenge with certification was trying to figure out how to coordinate and layer the various plans as they were revised to meet certification requirements.

Public Sector Responsiveness

Finally, there was the significant challenge of meeting the tight certification timelines. Rather than the typical long deliberation and gradual delivery of public forest management procedures and programs, certification required quick decisions and fast implementation. State governments are not as flexible as companies, and public land management requires greater deliberation in order to achieve a balance of forest values. As described by the Michigan DNR, there was discomfort with the pace of certification implementation: "Our implementation was hugely fast-tracked. With a tight implementation timeline some decisions had to be made very quickly and this was uncomfortable for the Department, as well as some constituents. When looking at natural resource policy a lot has to be deliberated and developed over time, not based on snap decisions."[47]

In some cases, a state, such as Washington, delayed certification in order to respect state forest processes that were underway. In other instances, the state worked with the auditors to accommodate the public land requirements. The Wisconsin Department of Natural Resources explained: "Our state was one of the first governments to get certified so there was some auditor learning required. Governments aren't able to respond as flexibly or quickly as private industry to some CARs. Our auditors were responsive to these concerns. As long as we were showing progress, then it was okay."[48] Minnesota also commented that its certification auditors had so far been responsive to public agency concerns and the different expectations of public forestland management.

Another challenge regarding public sector flexibility was the lack of budget allocation control. For example, explained the state forester of Maryland, public agencies not only may be tied to programs and procedures that prevent rapid certification response but also may be subject to budget, staffing, and management decisions that are made for reasons other than forest sustainability. According to New York's Division of Lands and Forests, the division entered into certification in good faith but did not control the budgets. Even though it was committed, it could not guarantee that there would be resources for follow-up. For example, during its certification implementation period, it lost several staff positions through attrition. The resource decisions were made centrally by a budget control officer and not at the departmental or divisional level.

Summary

As the previous examples have shown, state governments faced a range of challenges in implementing certification, including many administrative

Figure 5.12

Benefits of state forest certification

The bars indicate the extent to which the certified states identified each of the benefits. Interviewee responses were based on an open-question format rather than a structured survey. The breakdown of response by state is provided in the summary table in Appendix C.

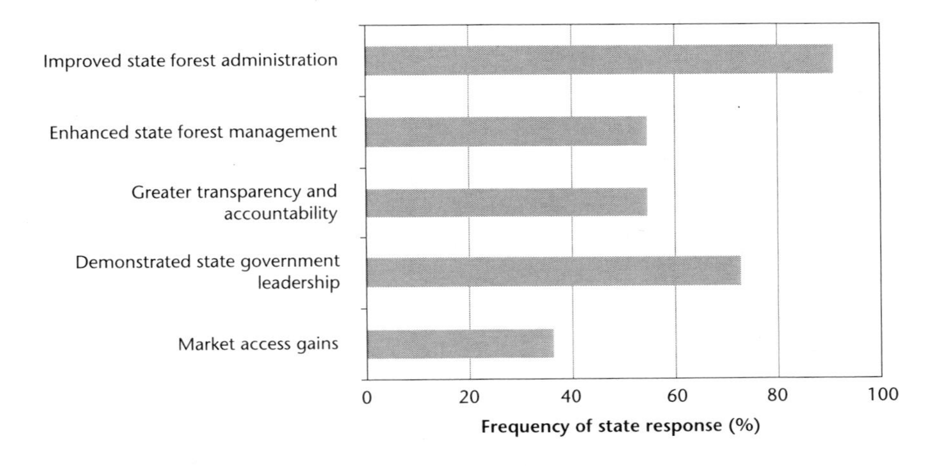

hurdles: staff resistance, managing increased workloads, justifying budgets, coordinating inter- and intradepartmental programs, and following the required state forest management and financial processes while also meeting the shorter certification timelines. Efforts to overcome many of these challenges translated directly into significant benefits, however, including improved state forest administration, better forest management practices, and an overall enhancement of state forest governance.

The Benefits of State Forest Certification

In my interviews with forest agencies in the certified states, I asked not only about the challenges but also about the positive and negative outcomes from certifying their state forests. It was an open question and interviewee responses were favourable. None of the states mentioned any negative outcomes, although there were some uncertainties about the future market benefits, the potential ratcheting of certification requirements, and the divided ENGO support.[49] Figure 5.12 summarizes the responses. The five most commonly identified benefit areas were improved forest administration, enhanced state forest management, greater transparency and accountability, demonstrated state government leadership, and some market access gains.

Improved State Forest Administration

The most common benefit of state forest certification across all of the certified states was improved state forest administration. Administrative issues that presented a challenge to implementation – difficulty in justifying costs, coping with staff reluctance, and coordinating certification across various governmental groups – ended up being key areas of benefit. Specifically, the certification of state forests helped state agencies improve staff morale, achieve greater program consistency and cooperative effort between departments and within divisions, increase process efficiencies, and leverage additional public resources.

Improved staff morale. Despite the initial reluctance of state forest agency staff to participate in certification, certification led to improved staff morale in many cases. Staff found that certification audits, rather than criticizing or dictating forest practices, were a means of recognizing and commending departmental efforts, providing constructive feedback, and identifying opportunities for improvement. In fact, some states referred to certification as a "morale booster" and a means to develop "organizational esteem" by achieving and maintaining certification under third-party oversight.

Better departmental coordination. Although the lack of inter- and intra-departmental coordination presented a challenge in implementing certification requirements, many states saw significant improvement in the consistency of their forestry division operations and a much greater degree of coordination among staff and specialists across departments. Several states commented that instead of separate, idiosyncratic governmental efforts, certification helped to establish an organization-wide approach to several key SFM issues, such as forests with high conservation values, rare and endangered species, and forest reserves. In other words, certification helped to keep everyone on the same page. It also helped bring together the wealth of specialists within government who contribute to state forest management, including ecologists, wildlife biologists, land surveyors, archeologists, and so on. The Massachusetts Bureau of Forestry even noted that by encouraging improvements in landscape-level planning, certification was helping the state coordinate its activities with forest owners on adjacent and nearby properties.

Greater administrative efficiency. For many states, certification acted as a springboard for addressing lingering issues and encouraged the more timely execution of state forest management responsibilities. Greater efficiencies were achieved through the implementation of more formal plan-do-check feedback systems, which included timelines and reporting to better track and ensure the completion of tasks. States commented that they were more organized as a result of certification and that the department was now doing what it said it would be doing. In particular, things had to get done to meet the annual certification surveillance audit requirements. Overall, the states commented that certification helped focus and accelerate their departmental activities and prompted greater efforts to "speed things along." For example, one state explained that whereas it had historically taken them many years to complete a management plan, certification introduced improvements that, even with more public input, reduced the process from twelve years to two years.

Improved access to state funds. For several states, a largely unexpected benefit of certification was that it gave their forest agency greater leverage in obtaining program funding that it had been trying for years to secure. For example, the Wisconsin Department of Natural Resources explained that whereas historically its requests for money for road maintenance had been turned down by the legislature, when the department received the audit corrective action request to address roads, the governor's office and legislature suddenly became supportive and provided additional base funding. The result was better management of roads and mitigation of environmental impact. In other words, by having a third-party independent auditor identify the need for resources, and with underlying political support for certification, state forest agencies had the justification they needed to obtain the necessary funds to update and/or carry out their forest plans and programs and meet their overall forest management responsibilities and objectives. State agencies compared having an independent certification audit in hand with having more ammunition to get additional resources. Legislators could clearly see what was required for the agency to do a good job, and there were new risks in not responding. In short, failing to meet and/or maintain the state's certification commitment was not a political option.

Enhanced State Forest Management

By adopting certification, state governments realized not just administrative improvements but also enhancements in their management of state forests, including improved technological resources and continual improvement in state forest planning, procedures, and forestry practices.

Updated technical resources. In response to certification requirements, many state agencies established new inventory systems, better forest models, and more formal monitoring, tracking, and reporting programs. Certification acted as the catalyst for developing and implementing this new technical capacity, which in turn led to gains in administrative efficiency and improved forest management. For example, having updated models and improved geographic information system (GIS)- and global positioning system (GPS)-based inventory enhanced governments' ability to identify endangered plants, pinpoint special management sites, track harvest by individual species, and better prioritize their work. As the Michigan DNR explained, "with our new monitoring, tracking and reporting systems, we can now document sites with environmental damage and needing some attention ... and better allocate funds to mitigation and management."[50]

Continual improvement in forest management. Forest certification also promoted continual improvement in the management of state forests. It encouraged forest agencies to question their forest management assumptions, better identify problems, refine their forest practices, and, overall, "take a deeper look at things." The states described certification as "shaking up their internalized feedback loops," requiring them to test their current forest management strategies. The specific benefit of the peer-to-peer review that certification enabled through regular independent forest audits was also mentioned – district foresters from other regions would tag along on audits and ask questions such as, "Why did you leave that tree ... that's not what we would have done." This promoted continual learning. Many states emphasized that they could not help but benefit from third-party assessment, that there was always room for improvement.

Overall, certification instilled and/or reinforced in agencies a culture of continual forest management improvement. The need to be prepared for annual audits forced bureaucrats to work on many facets of forest management with more emphasis than without any independent scrutiny. For many states, certification requirements and audit findings became a blueprint for further improvement. As the North Carolina Division of Forests explained, "certification provides additional structure to the process of properly managing a forest."[51]

Greater Accountability and Transparency

Forest certification also encouraged greater transparency and accountability through increased public engagement and the provision of more detailed forest information. The requirement for public consultation and regular tracking and reporting generally encouraged more positive public and ENGO feedback. Through the certification process, citizens had more opportunity to provide input and gain an understanding of the state forest objectives, challenges, and outcomes, and this fostered greater trust in state forest management practices. The availability of more detailed forest management information benefited not just the public but also industry and policy makers.

Demonstrated State Government Leadership

Several certified states, particularly those that certified early on, increased their profile and were seen as leaders in terms of setting an example and having the knowledge and expertise to assist their private forestland owners with certification.

For example, when New York achieved FSC certification, the local environmental and consumer groups took out a full-page ad in the *New York Times* praising the governor for certifying the state forests. Pennsylvania gained unexpected prominence at international trade shows, as well as special recognition within the US National Association of State Foresters. Certification also helped the states play a leadership role with their private industrial and non-industrial landowners where certification was concerned. For example, the Pennsylvania Department of Conservation and Natural Resources commented that whereas previously the state had benefited from industry's sharing of its experience, now representatives of industry were coming to the state for certification advice. In addition, faced with the challenge of encouraging the many small family forest owners to take a more active forest management role, state forest agencies could communicate more knowledgeably and convincingly about the certification process and opportunities. Certification also helped state governments compare and evaluate their leadership position. It provided a useful benchmark for policy makers in weighing the state's certification performance relative to other competing jurisdictions as well as to its own goals.

Market Access Gains

Although several states realized some market benefits from certification, for others the market gains were uncertain – either too difficult to measure or not yet realized. The three most commonly identified market benefits were aiding in the sale of state public timber, and therefore supporting state forest harvest levels and sustaining state revenues; helping industry in the state remain competitive and meet growing customer demands; and encouraging the in-state use of certified state fibre.

Although no state was receiving a market premium for its certified wood, in some cases the certified public fibre was given preference by local manufacturers at their "mill gate." State certification was, therefore, helping to maintain a local (non-export) market for unprocessed state timber as well as providing an incentive for local mills to chain-of-custody certify their operations and gain potential market benefits from selling certified forest and paper products. Several states commented that certification market demand was still developing and evolving, as certain sectors, such as solid wood manufacturing, were not fully on-board and chain-of-custody certifications were only just beginning to increase. In states such as Maine, the government had begun to shift from a supply-side focus on increasing the acreage of certified forest to a "demand-pull" strategy involving market campaigns and education to increase in-state market demand.[52] Overall, the market benefits of state certification were deemed to be just starting to emerge, with the states expecting to see more measurable market gains over the next few years as demand increased.

Overview of Co-Regulation Benefits

The foregoing examples illustrate that the co-regulation of state-owned forestland through certification facilitated a marked improvement in state forest governance capabilities, policies, and practices. In many cases, state forest agency resources increased, departmental coordination improved, staff morale improved, and greater organizational esteem was achieved. Enhanced public engagement and reporting increased the transparency and accountability of forest management processes, and improvements in state forest practices were realized through more formal tracking and monitoring systems, better models, updated inventories, and continual learning through independent third-party auditor feedback. Appendix D summarizes the range of positive forest governance outcomes (by state) as reported in the various state forest certification audit reports.

Summary

Returning to the opening puzzle, state governments in the United States responded directly and enthusiastically to forest certification as a result of a combination of economic, environmental, social, and political pressures as well as expectations of market gains and forest management improvements. Although market benefits remain uncertain, the trend towards state government certification co-regulation has been reinforced by positive forest governance outcomes.

The US forest industry is the largest in the world, and the administration of forests across the country is characterized by tremendous multilevel complexity, including variability in state-level forest tenure and regulatory arrangements across the country. Acknowledging the central role that state governments play in the delivery of private forest laws and the management of state forests, certification represents a notable trend towards a direct government co-regulatory role in forest certification. This trend has continued, with Indiana achieving certification in early 2007, Ohio committing to certification in late 2007, and several other states keeping a watchful eye on market developments in preparation for certification.

States were spurred to certify by a combination of market and non-market issue-based and opportunity-based drivers that played out differently across the various forest regions. Customer pressure, ENGO advocacy, declining state economies, and interstate competition were all *issues* that prompted state certification response. Private foundation funding, increased market access, and state government leadership were potential *opportunities* that influenced state governments to certify.

The timing of state certification was influenced not only by the combination of various drivers but also by the extent of debate over three questions: whether the state could afford it, which standard to pursue, and whether certification would threaten state forest policy authority. Certification proceeded as the states determined that the economic risks of not certifying were greater than the costs; that dual certification would satisfy the greatest constituency for minimal incremental effort; and that certification would complement state authority by verifying, reinforcing, and rewarding the state's sound forest practices.

Finally, the resolution of challenges to certification implementation served as a springboard for achieving positive forest governance outcomes such as greater administrative efficiencies and enhanced state forest management programs and practices. Specific benefits of certification included improvements in staff morale, departmental coordination, access to state funds, forest planning time frames, technological resources, forest management practices, public engagement, forest tracking and reporting, and state government forest management leadership. While there have been some market benefits in terms of increased local mill access for certified state fibre, measurable market gains have yet to fully materialize. In other words, state certification outcomes thus far have pertained more to forest governance improvements than to market benefits.

Overall, rather than being a substitute for the traditional role of state forest agencies in managing public forests, adoption of certification in state-owned forests is proving to be an innovative tool to supplement rather than supplant state government authority and enhance state forest governance capacity across the US. The dynamic interplay of public and private rulemaking systems encourages governance improvements, including policy innovation, the topic of the next chapter.

6
Sweden: Public/Private Forest Policy Interplay and Innovation

Sweden is a global environmental leader. The Swedish government hosted the first United Nations international conference on environmental issues in Stockholm in 1972, and the country consistently scores in the top tier in global environmental sustainability rankings. Sweden has also been a unique leader in forest certification. It was the first country in the world to develop a national Forest Stewardship Council (FSC) standard (in 1998) and was an initiator and founding member of the Programme for the Endorsement of Forest Certification (PEFC). What is particularly interesting about forest certification in Sweden is that it emerged in the early 1990s, around the same time that the government introduced a major shift in the country's forest regime from prescriptive regulations emphasizing timber production to more flexible results-based legislation that set broad environmental goals. This coincidence raises questions about whether certification stepped in to fill a regulatory gap. Did the state retreat and private rule making in the form of certification substitute for traditional public forest policy authority in Sweden?

This chapter examines the dynamic interaction between the public and private forest governance systems in Sweden over the past fifteen years. Drawing on my in-depth interviews with Swedish certification experts and practitioners in 2007 (see Appendix A) as well as the secondary literature on certification in Sweden,[1] the chapter reveals how certification and public policy have operated in parallel. The public and private forest rules have been contested back and forth and continually improved. Rather than retreating, Swedish forest authorities recognized both the opportunities and limitations of certification and thus continually revised public forest policies and regulations while enabling and leveraging forest certification as an additional governance tool within the government's overall policy mix. Specifically, I argue that the certification/policy interplay in Sweden contributed to a sharpening of the national forest goals and targets, enhancements in the sustainable forestry discourse, and an ongoing testing and strengthening of the country's sustainable forest management (SFM) vision.

The chapter begins with an overview of forestry in Sweden, consisting of a snapshot of the development and evolution of Swedish forest certification and the expectations and positioning of the government's role in certification. In particular, I explain how the Swedish government provided an enabling policy climate for certification development, implementation, and ongoing improvement. I then examine the interaction of the public and private forest governance systems and argue three central points. First, the timing of the establishment of a new forest policy regime in Sweden created a window of opportunity for co-regulatory governance. Second, a dynamic interplay between certification and public policy helped to define and continually improve on the national forest objectives and targets. And finally, as analyzed in the last section of this chapter, certification added to the forest discourse in Sweden by creating a new forum for SFM policy engagement and by challenging forest owners to define and operationalize the national vision of balancing forest production and environmental goals.

The Swedish Forest Regime

Sweden is a heavily forested Nordic country with a large industrial forest sector. With less than 1 percent of the world's forestland, Sweden is the world's third-largest exporter of sawn timber and pulp and the fourth largest exporter of paper. Sweden is a major player in the global timber economy and has been an international forest certification leader, accounting over a recent decade (1995-2005) for just under 10 percent of the export value of globally traded forest products and just over 10 percent of the global certified forest area. Five main features characterize forestry in Sweden and have contributed to the country's high acceptance of forest certification:

- intensively managed secondary-growth boreal forest
- a majority of family-owned forest concentrated in the south
- a highly fragmented fibre supply
- dependence on exports to the European Union (EU)
- results-based forest legislation that aims to balance environment and production forest values.

Intensively Managed Secondary-Growth Boreal Forest

The Swedish landscape is largely boreal and is therefore dominated by coniferous forests, with Norwegian spruce and Scots pine constituting roughly 85 percent of the timber stock. Deciduous forests (mainly birch and aspen) account for approximately 15 percent of the country's 28 million hectares of productive forestland. Swedish forests have been managed since the eighteenth century and nearly all forests have been influenced by human

Figure 6.1

Forest ownership in Sweden

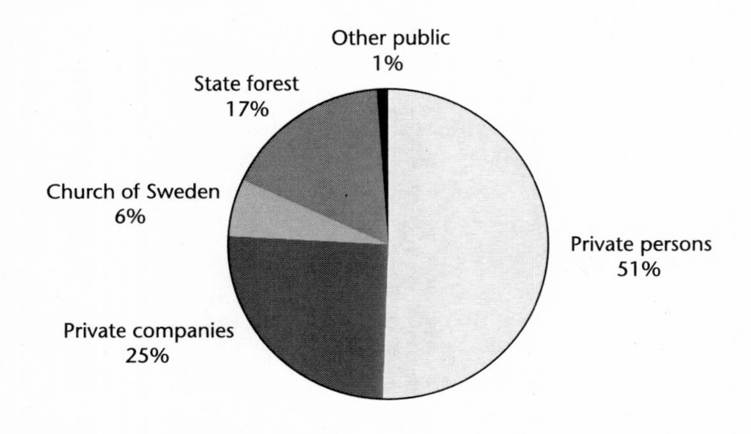

Source: Swedish Forest Agency 2007, 317.

activity. Only 4 percent of Swedish forestland is characterized as "old natural forest." These stands are found mostly in the northern interior regions of the country and are largely protected in national parks and nature reserves.

A consequence of production-oriented forest regulations and intensive silviculture practices over the past half-century is that Sweden achieved the highest level of forest productivity of any northern timber-producing nation but ecological forest values suffered losses as a result of the emphasis on production.[2] In order to address these losses, nature conservation and increasing forest biodiversity are now also Swedish forest policy priorities. Forest certification has provided a complementary mechanism to assist in achieving these broader sustainability forest objectives.

Family Forest Ownership

As shown in Figure 6.1, 82 percent of Sweden's forests are privately owned by individual families, companies, and the Church of Sweden. Families make up the largest single category of forest owner (51 percent), followed by private forest companies (25 percent). There are approximately 350,000 individual family forest owners who manage many small forest properties across Sweden, totalling approximately 11.7 million hectares of productive forest. The Church of Sweden is also a large private forest owner, controlling approximately 1.4 million hectares (6 percent). Public forestland accounts for 18 percent, most of which is managed by the state-owned forest company, Sveaskog. The National Property Authority manages 882,000 hectares of the state-owned productive forest. Municipalities and county councils own and

Figure 6.2

Regions of Sweden

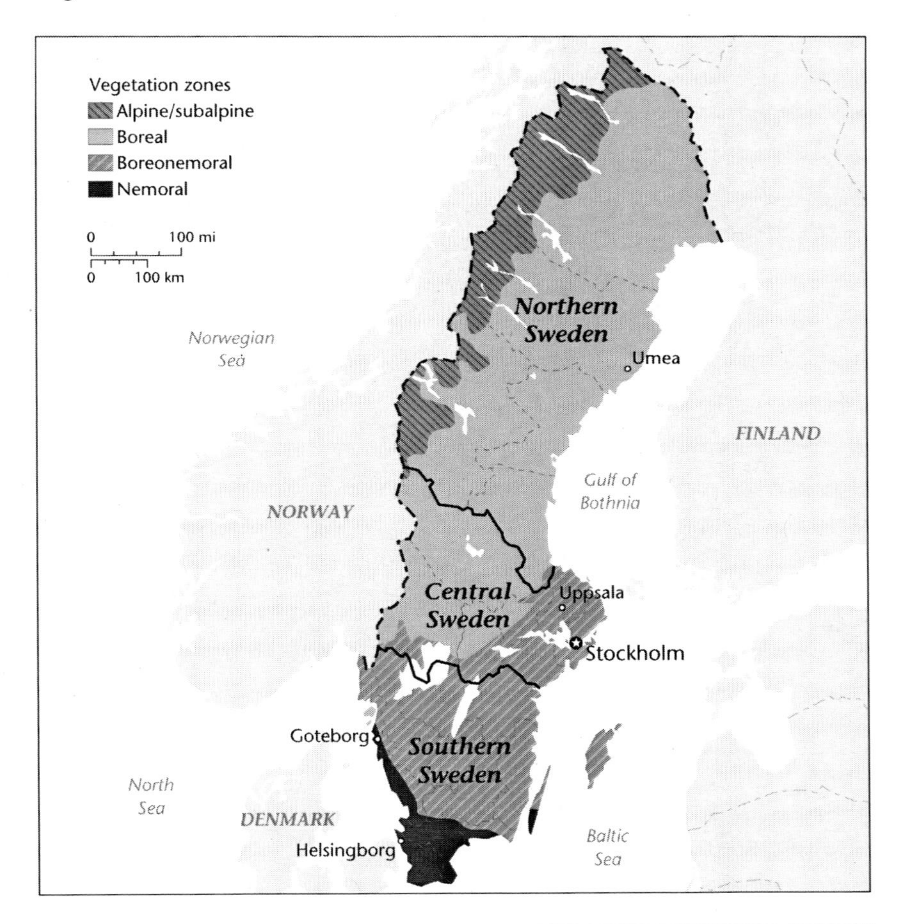

Source: Swedish Forest Agency, http://www.svo.se.

manage approximately 324,000 hectares, or just under 1 percent of Sweden's productive forests.

Although the vast majority of Swedish forestland is privately owned, there is a common-law right of free access to all woodland that is not part of a private dwelling. This "right to roam" has helped to minimize land-use conflicts, particularly regarding shared recreational use of the forest.[3]

Sweden is divided into three regions: the North (Norrland), Central (Svealand), and South (Götaland) (Figure 6.2). Most of the population and the most productive forests are located in the South. Eighty percent of the forestland in the South is family-owned, compared with 35 percent in the North. The North is sparsely populated and includes mountainous alpine and subalpine regions on the west coast. Private forest company land is located primarily in North and Central Sweden. Sveaskog state forest holdings are distributed throughout Sweden, with the majority in Norrbotten and Västerbotten counties in the North. The North is also characterized by traditional Sámi territory and forestlands for reindeer grazing. Southern forests have the highest productivity (including boreonemoral and nemoral forests),[4] but have also historically been the most intensively harvested and/or altered. Over the past few decades, however, the forest area in the South has been increasing, largely because of the conversion of low-yielding agricultural land back to forest in response to the increased financial value of forests as a raw material input to the forest industry. Overall, forest owner SFM interests and concerns differ between the northern and southern regions of the country, and this has been reflected in the country's certification debates and the specific regional adaptations within the Swedish certification systems.

Fragmented Fibre Supply

The fibre supply in Sweden is highly fragmented and the demand is increasingly competitive. Although the large Swedish forest companies are integrated (operating mills as well as having access to sizable land holdings), they also rely on timber purchases from the many individual private forests across the country.[5] Currently, family forest owners account for 60 percent of Sweden's timber production. The average size of a family forest in Sweden is only fifty hectares, so to coordinate production and help achieve economic efficiencies, family forest owners are encouraged to join one of Sweden's four main private forest landowner associations under the Swedish Federation of Family Forest Owners (Skogsägarna LRF).[6] Sweden's private owners are very independent, however, and less than 50 percent have chosen to be LRF members. Securing access to fibre thus remains competitive, dynamic, and an ongoing challenge.[7] The fragmentation of the fibre supply has been a central issue in designing feasible chain-of-custody forest certification systems for the country.

Pulp mills and sawmills are distributed throughout Sweden, with over half of the sawmills located in the South. Of the approximately 86 million cubic metres of fibre consumed annually within the country, pulp and paper mills utilize 47.6 million cubic metres (55 percent) and sawmills 37.2 million cubic metres (43 percent). Most Swedish sawmills are independent (not owned by the private landowner associations or forest companies) and account for roughly 65 percent of the country's solid wood production. The sawmilling industry in Sweden has become increasingly centralized as a result of closures and consolidations caused by the economic downturn in the sector in the late 1990s. Most of the independent private sawmills are members of the National Federation of Private Independent Sawmills (Sågverkens Riksförbund).[8] Setra is Sweden's largest sawmilling company, with ten to twelve sawmills across the country; it is 50 percent state-owned by Sveaskog and 26 percent by Mellanskog LRF. The private sawmills are of note as they played a leading role in the establishment of PEFC certification in Sweden; they supported the PEFC standard largely because of operational concerns regarding the feasibility of meeting the FSC's initial chain-of-custody fibre segregation and tracking requirements.

EU Export Dependence

Forestry is a major sector of the Swedish economy, accounting for 12 percent of the country's exports and 4 percent of GDP. The industry is dependent on exports, with about 70 percent of sawn wood products and 80 percent of paper production going to Western European markets. For example, over the past decade, Sweden has supplied European customers with approximately 10 percent of their paper needs and 12 percent of their sawn lumber demand. Germany and the United Kingdom are Sweden's biggest export customers and were very influential in driving Swedish companies to seek FSC certification. In order to meet their fibre demands and production requirements, Swedish forest companies also rely on fibre imports (particularly deciduous pulpwood) from Russia, Latvia, Estonia, and Norway.[9] A recent challenge for the Swedish government has been to control the importation of illegal timber.[10]

The New Forest Regime: "Freedom with Responsibility"

The final distinguishing feature of the Swedish forest regime is the country's long history of prescriptive forest management enforced through information, outreach, and moral suasion rather than punitive penalty. Over the past decade, Swedish forest policy has also undergone a major transformation, from policies based on maximizing sustained yield to multiple-use forest legislation promoting the balancing of environmental conservation and timber production goals.

Forestry in Sweden is regulated nationally under two main pieces of legislation: the Forest Act (1994) and the Environmental Code (1998). The Forest Act sets out the conditions for timber harvesting, forest regeneration, maintenance of forest health, and protection of cultural and environmental forest values. The Environmental Code sets out the requirements relating to the conservation of ecological values, including the protection of habitat for endangered species. Although not formally linked to the Forest Act or the Environmental Code, specific environmental targets for sustainable forests are also defined under the national "Sustainable Forests" objective established by the Swedish parliament in 1999 (Table 6.1).[11] The purpose of this objective

Table 6.1

Swedish Sustainable Forestry objective and national targets

Interim targets	Requirements and time frame
Target 1 – Long-term protection of forest land	From the base point of 1998, a further 900,000 ha of high conservation value forestland will be excluded from forest production by the year 2010.
Target 2 – Enhanced biological diversity	By 2010, the amount of dead wood, the area of mature forest with a large deciduous element, and the area of old forest will be maintained and increased by: • Increasing the quantity of hard dead wood by at least 40% throughout the country and considerably more in areas where biological diversity is particularly at risk. • Increasing the area of mature forest with a large deciduous element by at least 10%. • Increasing the area of old forest by at least 5%. • Increasing the area regenerated with deciduous forest.
Target 3 – Protection of cultural heritage	By 2010, forestland will be managed in such a way as to avoid damage to ancient monuments and to ensure that damage to other known valuable cultural remains is negligible.
Target 4 – Action programs for threatened species	By 2005, action programs will have been prepared and introduced for threatened species that are in need of targeted measures.

Source: Swedish Forest Agency 2005.

is to maintain the functionality of ecosystems, preserve the natural biodiversity of Swedish forests, and safeguard their cultural heritage and other social values.

Forest policy development has fallen alternately under the authority of the Ministry of Agriculture or the Ministry of Industry. As of 1 June 2007 (the time of this study), it resided in the former. Forest policy implementation is the responsibility of the Swedish Forest Agency (SFA), which before January 2006 was referred to as the National Board of Forestry (NBF). The main role of the SFA is to supervise compliance with the Forest Act and the Environmental Code. The SFA oversees 120 local offices, whose primary role is to deliver national forest policy by providing forest management advice and information to private forest owners. Besides providing extension services and advice to private forest owners, the SFA gathers and publishes statistics about the forest sector, carries out annual inventories, monitors forest health, conducts forest management planning, and, in cooperation with the Swedish Environmental Protection Agency (under the Ministry of Environment) and County Administrative Boards, supports nature conservation efforts.

One of the major drivers of forest regulation in Sweden has been fear of a national fibre shortage. As far back as the mid-nineteenth century, the Swedish government recognized that agricultural conversion and growing industrial demands were depleting the nation's forests. As a result, since the first Forest Act in 1903, forest policy has emphasized forest regeneration and production. The forest industry and government have historically worked very closely and cooperatively to ensure a sustained timber yield. The policy focus changed in the early 1990s, however, as a result of increased public concern and awareness regarding nature conservation and the protection of non-timber forest values.

Following a national forestry commission in the late 1980s, the Swedish Forest Agency introduced a new Forest Act in 1994 and brought about two major transformations in the Swedish forest regime: (1) Sweden's forest policy focus changed from a historical emphasis on production to balancing of environmental and economic forest values; and (2) the government's policy approach shifted from prescriptive regulation to results-based legislation. According to the revised Forest Act, "forest should be sustainably managed aiming at ensuring the production of high and valuable yield and at the same time ensuring forest biological diversity and the possibility of multiple uses of forest, now and for future generations." Rather than spell out specific operational-level regulations, the Forest Act was frame law (principles rather than prescriptive rules) based on the premise of letting landowners decide how best to achieve a balance between the environment and productions goals. The government described its approach as "freedom with responsibility." In order for landowners to maintain their freedom under the new regime, they needed to demonstrate their SFM responsibility. This involved first determining the optimal desired balance of forest values and then implementing the most appropriate SFM practices. The government would closely monitor progress, and would introduce appropriate regulations if it found that either production or environmental values were suffering. As we shall see, the government's new forest policy approach, along with the industry's desire to fend off market protests led by environmental nongovernmental organizations (ENGOs), created favourable conditions for the development and adoption of forest certification in Sweden.

Sweden's Forest Certification Leadership

Sweden accounts for 6 percent of the world's certified forest. The country has been a global leader in the development and adoption of FSC certification.

Figure 6.3

Certified forest in Sweden by ownership category, 2007

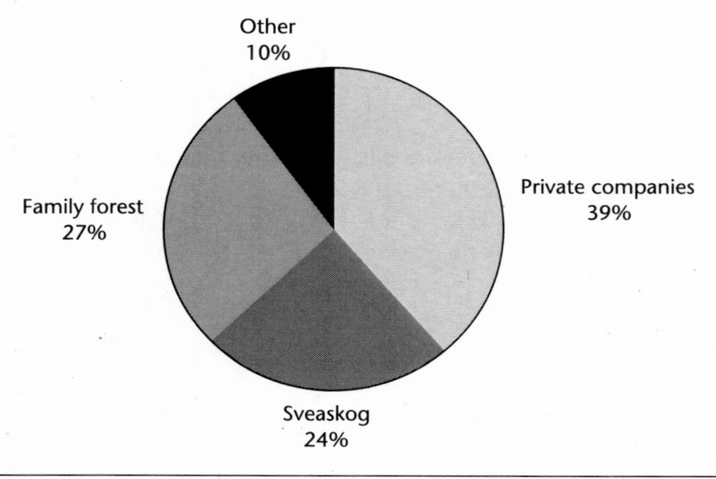

Sources: Svenska PEFC, http://www.pefc.se; Svenska FSC, http://www.fsc-sverige.org.

It was the first country to establish an FSC national standard and currently ranks third after Canada and Russia for total area of FSC-certified forest. The state-owned company Sveaskog holds the second-largest FSC certification in the world (3.4 million hectares).

Certification Status

As of late 2007, Sweden had certified over 60 percent of its productive forestland – approximately 7.4 million hectares to the PEFC standard and 10.4 million hectares to the Swedish national FSC standard (some forest is dual-certified).

Most family forests have been certified under the PEFC standard and forest companies account for most of the FSC-certified forest (Figure 6.3). The largest private company certifications include Svenska Cellulosa Aktiebolaget (SCA) (2 million hectares), Bergvik (1.9 million hectares), and Holmen (1 million hectares). An increasing number of companies (including Bergvik and Holmen) have also dual-certified to the PEFC standard. The Church of Sweden has been divided in its support – eight parishes have certified to the PEFC and five to the FSC standard. The state forests have certified to the FSC standard only.[12]

Certification Development and Adoption

As noted previously, there are two certification standards operating in Sweden – the FSC and the PEFC – and Sweden has been a leader in the development of both. The Swedish FSC national standard was established in 1998 and the Swedish PEFC standard was endorsed in 2000.

Certification Development

Forest certification emerged in Sweden following a range of international forest controversies within the country in the 1980s, including the clear-cutting of old-growth forest in the North, chlorine pulp bleaching, and the use of non-indigenous species. In 1994, WWF Sweden, in cooperation with the Swedish Society for Nature Conservation (SSNC), established an informal group to begin drafting a Swedish FSC standard. By January 1996, the Swedish forest industry became engaged in the process, largely because they were getting direct market signals from their large Western European customers, who were under intense ENGO advocacy pressure to source FSC-certified forest and paper products. As well, Swedish companies were acutely aware of the market influence of international advocacy groups, as they had experienced the effective ENGO-driven anti-chlorine bleaching campaigns in the 1980s. And finally, the industry's attempt to develop a competing Nordic certification standard had failed.[13] Unlike forest companies in North America, which adopted a defensive stance towards the FSC (out of fear of harvest reductions), Swedish forest companies therefore took a proactive position, advancing the development of a national FSC standard to quell ENGO pressure and give their industry a potential competitive advantage in the global market.

The Swedish FSC national working group was formally established in February 1996 and within fourteen months had reached agreement on a draft national standard.[14] The working group submitted the standard to FSC International in the fall of 1997. It was ratified on 26 January 1998 and published on 5 May. Although the working group reached consensus surprisingly quickly, it achieved this without resolving certain fundamental SFM questions and without the support of two major groups, Greenpeace and the Swedish Federation of Family Forest Owners.[15]

Greenpeace withdrew from the FSC working group early in the process, considering the discussions with industry a sell-out over the permission of intensive industrial forest methods (clearcutting, the use of exotic species, fertilizer and pesticide use, and so on). The family forest associations pulled out in April 1997, deeming the standard to be biased towards large industrial operators and not sufficiently reflective of family forest owner interests as per the structure and operations of small privately owned forests.[16] They also knew that they were in an excellent position to establish their own standard because several landowner associations had already developed their own environmental management standards. The family forest owner associations and the private independent sawmills of Sweden began working immediately to formalize an alternative internationally recognized national certification program that provided greater flexibility to suit the range of private forest owners.[17] The Swedish PEFC Interim Council held its first meeting on 23 June 1999, and within six months had completed a draft of the Swedish PEFC standard, which included the group certification standards of the various Swedish landowner associations.[18] The standard was endorsed under the international PEFC program in May 2000.

In autumn 2000, the Swedish FSC and PEFC programs attempted to address some of the logistical implementation problems that were the result of having two separate standards, by creating a bridging document to mutually recognize the two standards. The cooperative effort was referred to as the Stock Dove process *(skogsdovan),* referring to the attempt to bring peace to the forest and decisions over the timber stock. Although there were meetings of the various stakeholders and the Stock Dove committee developed a set of recommendations by December 2001 to harmonize the two standards, the process never went forward.[19] With the exception of the requirement to give Sámi greater access to private forest, however, the Swedish PEFC did incorporate the Stock Dove recommendations (2002 and 2004 revisions) in order to close the gaps between the standards.[20] The FSC, on the other hand, wanted to maintain the rigour, independence, and distinction of its standard from the PEFC program, and did not follow up on the Stock Dove recommendations.

FSC Sweden commenced its standard revision process in 2003, and after intensive debate over key SFM issues that had been left unresolved in the first version of the standard, the groups finally reached agreement on a revised draft standard in May 2005. FSC International rejected the standard, however, as it was deemed not specific enough. Two more years of debate produced another draft, which FSC International also rejected in the fall of 2007. Most stakeholders remain engaged in the continuing revision process, but with a high level of frustration. Early in 2008, Sweden's lead ENGO – the SSNC withdrew from the Swedish FSC board, explaining that the standard was weak and "the lack of observance substantial."

Despite the failure of the Stock Dove process to bring about mutual recognition of the PEFC and FSC standards in Sweden, the standards have nevertheless become closely harmonized. For example, an independent consultant's comparative study of the Swedish PEFC and FSC standards conducted in 2005 found that "there are not any essential differences between the requirements of the two standards apart from the slightly higher harvesting restrictions in FSC certification and stricter commitment by contractors in PEFC certification" (Savcor 2005, 70).

Certification Adoption

Certification uptake in Sweden occurred early and rapidly (Figure 6.4). All the major forest companies were certified to the FSC standard by 2000; in

Figure 6.4

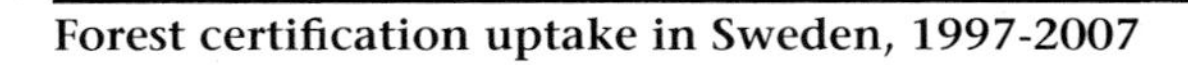

Forest certification uptake in Sweden, 1997-2007

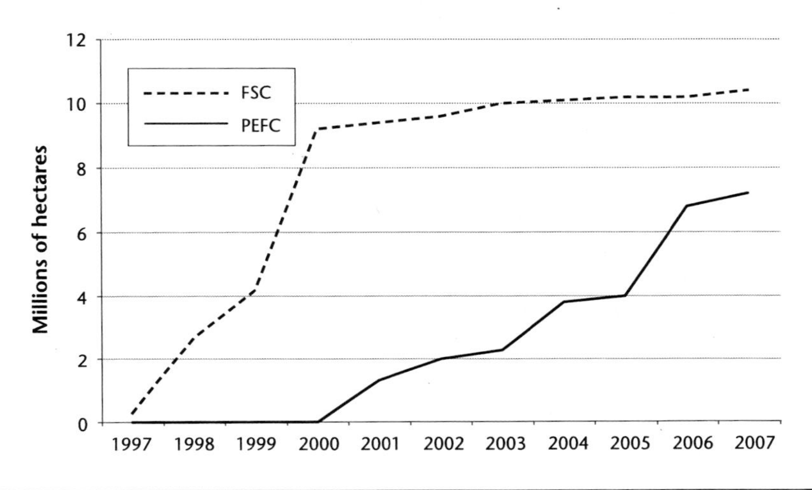

Source: Svenska PEFC, http://www.pefc.se; Svenska FSC, http://www.fsc-sverige.org.

fact, many had FSC-certified even before the FSC board approved the Swedish standard.[21] The first PEFC certifications also preceded formal international approval of the standard in 2000. Several parishes, private independent sawmills, and the landowner associations were all early PEFC adopters. Following the initial certifications, there was an increasing trend among the private independent sawmills and the large forest companies to dual-certify in order to facilitate chain-of-custody certification of their fibre supply and have enough production flexibility to meet customer demands. For example, Holmen certified 1.275 million hectares and Bergvik 2.3 million hectares to the PEFC standard in 2003 and 2004, respectively.

Many family forest owners adopted the PEFC not only because the requirements were less financially costly (and therefore more suited) to the small forestland owner than the FSC but also because the landowner associations offered assistance for participating in PEFC group certifications. And as mentioned previously, a significant factor leading the Swedish forest companies to embrace the FSC (rather than the PEFC) was their first-hand knowledge and fear of the powerful influence of ENGO market campaigns and the resulting pressures to FSC-certify coming from their large UK and German customers.

Beyond the specific factors that influenced the *type* of certification system adopted, an underlying certification driver in Sweden was the desire of all forest owners to support the government's newly introduced results-based forest policy approach by demonstrating their voluntary SFM commitment.

Forest certification provided an opportune vehicle. For example, the landowner associations argued convincingly to many of their members that to maintain regulatory freedom, family forest owners had to demonstrate responsibility and that PEFC group certification was the appropriate and feasible means. Thus, certification in Sweden was driven not just by ENGO and market forces and the competition between the standards but also by the policy climate. As we shall see next, the state played an important role in enabling the participation of forest landowners in certification by placing direct expectations on voluntary SFM initiative.

The Swedish Government's Role in Certification

The Government's Position

The Swedish government's position on forest certification was to passively observe and not interfere in the certification dynamics. The Swedish Forest Agency supported certification as one tool among several to promote sustainable forestry and achieve national forest policy goals, but did not formally engage in certification to drive the private rule-making governance process. Consistent with the European Union's position on certification, the Swedish government (including the Swedish Forest Agency and the Swedish Environmental Protection Agency [SEPA]) viewed certification as a market instrument. As explained by the State Forest Agency: "Whether a red or blue government, certification is a market-driven process ... and if a market instrument is working then government shouldn't interfere."[22] SEPA clarified: "We don't really know about what is going on with certification because we were told that authorities should not be involved – we do not go into the certification working groups. But government is very supportive of the certification process."[23]

Expectations of Government's Role

During my interviews with Swedish forestry stakeholders in the fall of 2007 (Appendix A), I asked them to describe the position and role that the Swedish government had taken in certification, and whether or not this was appropriate. The responses were consistent. All stated that the government's position was one of non-interference and that this was exactly appropriate – that certification should not be state-driven. As the former chairman of the Swedish Federation of Family Forest Owners noted, "government has been where they should be – on the sideline."[24] The chief forester of this same federation concurred: "No one wanted government involved and they were reluctant to engage."[25] Government officials agreed: "No one argues that our position of observer is wrong – everyone thinks this is good."[26] Specifically, it was explained that one of the main reasons no one wanted government directly involved as a stakeholder in the certification process

Figure 6.5

The Swedish government's role in certification

Spectrum of government intervention

Indirect ——→ **Direct**

	Observe	*Cooperate*	***Enable***	*Endorse*
Standards development	✓	SFA and SEPA provided policy clarification and feedback on draft standards. FSA working group chair came from SEPA.	**Forestry frame law enabled private certification initiative.**	
Implementation	✓	Local authorities responded to landowner certification information requests.		Sveaskog leadership
Enforcement	✓		SWEDAC PEFC auditor accreditation	
	Government position	Government role		

was that as the legislator of forest policy, government was not an equal party and would distort the balance and overly influence decisions.[27] It was also pointed out that "forestry officials were under strict orders not to 'mess' with certification – that the *Forest Act* includes expectations but not specific rules. How to achieve these expectations shall be left to the forest owners."[28] In other words, government direct engagement in certification rule making would have been contrary to the legislative approach.

The Range of Government Roles in Certification

Although the Swedish government's approach towards certification was one of non-interference, it nevertheless contributed to forest certification at the development, implementation, and enforcement stages. These examples are summarized in Figure 6.5 and outlined below. In particular, Figure 6.5 highlights the most influential governmental role – introducing broad frame law forest legislation that provided by default an enabling policy environment for private certification development and implementation. The forest governance implications of certification within this forest policy context are the focus of the rest of this chapter.

Standards Development

The PEFC and FSC standards overlapped with government policy fundamentally because they required legal compliance. For example, the introduction to the Swedish PEFC standard states: "The principles, rules and guidelines contained in the Swedish Forestry Act and other relevant legislation constitute the basis of the Standard." Although government representatives were not invited and did not attend either the PEFC national council or the Swedish FSC working group standard development meetings, the working groups consulted the legislation and the forest authorities to ensure the alignment of the standards with government policy. The SFA and SEPA closely followed the proceedings and provided clarification of state forest policy during the development of the standards. For example, the SFA and the regional SEPA agencies cooperated by going on excursions with the Swedish FSC council to clarify the definition and implementation of requirements relating to key habitats. Early on, the FSC sent the agencies draft copies of the standard for their review and comment, calculated the effect of the certification standard on the level of harvests in Sweden, and shared this information with government authorities for their review.[29]

In addition, the government was able to closely follow the development of the FSC national standard as the FSC working group chose as their neutral chairperson SEPA director Lars-Erik Liljelund. Liljelund participated as a citizen rather than as a government official, and his role was clearly to facilitate rather than influence the process. SEPA, however, did provide his time free of charge to assist in the process, and this was viewed as government's indirect approval of certification. Both SEPA and the SFA were saying positive things about certification at the time; "they saw it as a complement to the legislation."[30]

Swedish authorities were also directly consulted during the PEFC standard revision process in 2004. The SFA, SEPA, and the Swedish Central Board of National Antiquities were invited to three meetings with the PEFC working group to discuss the revisions.

Implementation

The SFA's regional forestry boards (prior to the 2006 reorganization) and the government's county boards had knowledge of certification systems and requirements, and although these regional-level national administrative bodies were not formally engaged in certification implementation, the government employees did respond on an ad hoc basis to local forest owner information requests. When asked about the interaction, SEPA commented that if county board employees were engaging in certification, it would have been as individuals rather than as government representatives, as the county boards were under instruction to not interfere in certification.

Besides informal engagement in certification at the regional level, the government played an indirect role in endorsing certification by supporting the certification of the country's public land under the state-owned company Sveaskog (known as AssiDomän at the time). FSC certification of the state forests demonstrated the government's support for certification and provided an example of leadership in implementation for the many private forest owners.

Enforcement

The Swedish government has played a direct role in the enforcement of certification since the PEFC auditor accreditation process has been carried out by the Swedish Board for Accreditation and Conformity Assessment (SWEDAC), a public authority under the Ministry for Foreign Affairs.[31] Specifically, SWEDAC assessed the competence of the PEFC certification audit bodies. (The FSC accredits its own auditors independent of the state.) In addition, the SFA indirectly kept an eye on certification effectiveness through its annual Polytax inventory surveys. The SFA explained that if forest conditions were found to be deteriorating (for example, landowners were not matching their freedom in forest management with demonstrated responsibility), prescriptive legislation was always an option. Finally, the government considered the development of a public timber procurement policy to discourage illegal timber imports and encourage the domestic supply and government purchase of certified forest and paper products.[32]

In summary, although the Swedish authorities made an effort to stay out of the way of certification, they cooperated in the hope that the private governance system would provide a positive contribution to the overall goal of balancing production and environmental values in the forest. Specifically, SEPA and the SFA hoped that the detailed requirements of certification would complement the goals of the Healthy Forests national environmental objective and the guidelines of the 1994 Forest Act and the 1998 Environmental Code. By introducing frame law forest legislation, the government established a policy environment that favoured voluntary self-regulation and enabled certification development, adoption, and ongoing improvement. A multi-centric system of forest governance was established in Sweden as a result of the interaction between public forest policies and private certification rules.

Certification/Policy Interaction

Certification and forest policy in Sweden have shared a consistent broad vision of sustainable forest management in terms of balancing and sustaining economic, social, and environmental forest values. Their implementation timing has also coincided. The public and private governance systems have differed, however, with respect to decision-making processes and certain key

SFM definitions and forestry requirements. Thus, the systems have not simply been interchangeable. Instead there has been a certain degree of contest and back-and-forth interaction, with certification standards incorporating and going beyond state legislation, and state forest policy in turn advancing beyond the private rules. The dynamic interplay between private forest certification standards and public forest policy established a co-regulatory forest governance system in Sweden that has ultimately facilitated both greater SFM discourse and the continual improvement of the country's forest vision and SFM targets.

The Window of Co-Regulatory Opportunity

Certification emerged in Sweden in 1994, at the same time that the government introduced changes to the national forest regime. Through the 1994 Forest Act, the government delegated SFM self-regulatory responsibilities and created a window of opportunity for voluntary certification and the establishment of a co-regulatory forest governance approach in the country. As previously explained, forest certification offered not only a unique marketing instrument but also a pragmatic tool for companies and family forest landowners to define their SFM responsibilities under the broad forest legislation. Ultimately, certification was a way for forest owners to operationalize the general goals of the new frame law legislation and demonstrate their SFM commitment.

From the initial stages, certification and forest policy were mutually beneficial. The frame law forest legislation established a favourable context for certification development and implementation, and certification provided a means to assist in the delivery of the government's forest goals. The Swedish Forest Agency also recognized, however, the challenges and limitations of certification, and the fact that it was by no means a replacement for legislated forest rules. Certification was voluntary, and the definitions and targets did not precisely align with the government's forest objectives. Also, although certification offered an additional forest management accountability mechanism, it fell short in terms of providing a means to monitor and enforce requirements with respect to improving actual, on-the-ground forest conditions, particularly at the landscape level. Thus, with the development and increased uptake of certification among Swedish forest owners, the state did not retreat but rather facilitated, leveraged, and improved upon the accepted certification rules and mechanisms alongside state forest regulation.

While the timing of forest certification and the new forest policy regime coincided in the early 1990s, certification was *not* the driver of the government's change in policy approach and direction. The government did not redesign the national forest legislation in order to enable forest certification. Rather, the regulatory shift largely came out of the government's overall strategy to streamline government services, as well as in follow-up to the national forest commission discussions in the late 1980s, both of which occurred before the development of certification.[33] As a SEPA official explained, "I was involved in the *Forest Act* development and there was no discussion of certification. The revisions were made before certification – certification came later."[34] Several interviewees also noted, however, that the government was well aware of certification when it was designing the changes to the Forest Act. For example, the government knew when it set the voluntary forest reserve goal that certification would be there to possibly help to achieve the target. In other words, the agencies saw an opportunity for certification to serve as a potentially important regulatory complement to attaining the national environmental objectives.

Enhancing SFM Targets

To a large degree, certification did complement the new forest regime. The specific forest certification requirements provided a vehicle for translating the government's broad environment/production goals into specific operational targets and plans that landowners understood and could implement.

Although Sweden's forest owners welcomed the increased freedom under the 1994 Forest Act (with a high level of discretion to balance production and environmental values in their forests), the lack of specific, legislated SFM instructions also created operational and governance challenges. Many forest owners considered the national forest goals to be abstract and vague, and were uncertain of the government's precise expectations regarding how to translate the broad goals to optimal on-the-ground SFM practices. For example, as one interviewee explained, "As a forest owner you don't know what the benchmark is ... the government talks about an 'advisory level of operation' but there is no document saying what this is." As well, although the SFA conducted annual inventories and consistently reported that 25 percent of the harvest areas did not reach "the requirements," there were no fines or follow-up so forest owners did not know whether or what they had done wrong. There were also ambiguities in the forestry objectives. On the one hand, natural functions and processes of forest ecosystems were to be upheld; on the other hand, the natural processes of flooding and burning were to be controlled (Hysing and Olsson 2005). Generally, forest owners found the national forest policy goals and objectives to be overly vague general statements that lacked specifics, that is, a way forward.

Forest certification therefore supported the frame law forest legislation by offering forest owners specific SFM criteria and instruction – an "SFM pathway" to navigate the government's broad forest policy goals. The role of certification in enhancing SFM targets was particularly evident in three areas:

- operationalizing the forest goals through "green management plans"
- contributing to the national voluntary forest reserve target
- refining specific forest structure objectives.

Each of these examples also illustrates how the government recognized the need for state forest policy to be adaptive and evolving alongside certification's continually improving private rules.

Operationalizing the Forest Goals: Certification and Green Management Plans

"Green management plans" (forest plans with nature conservation considerations) were a clear example of certification's playing a significant co-regulatory role in Swedish forest governance. At the same time that government deregulated forest management plans in 1994, forest certification made long-term SFM planning a necessity by adding measurable targets and enforcement audits to the forest agency's forest management planning guidelines. Overall, by requiring management plans that incorporated environmental sustainability considerations in accordance with the government's green management plan guidelines, certification systems helped operationalize the national forest goal of balancing production and environmental forest values.

Before 1994, in order to ensure a sustained national timber supply, there was a requirement under the 1979 Forest Act that every private forest owner in Sweden have a forest plan describing the forest conditions and management suggestions. In support of the prescriptive legislation, county board foresters provided education and technical services to assist forest owners with plan development. Until 1994, therefore, most of the family forest owner management plans in Sweden were developed by public agencies. The 1994 Forest Act removed the mandatory forest plan requirement and cut back on SFA technical services. Although management plans became voluntary, the SFA realized that sustainable forest management plans were critical to achieving the national sustainable forestry goal of balancing environmental and production forest values. To ensure that environmental values were incorporated into management plans, and acknowledging the fact that certification systems were defining their own management plan criteria, the SFA began defining guidelines for *green* management plans. Rather than maximization of timber production, the goal of such plans was to manage for sustained production *and* nature conservation.

As mentioned earlier, the greening of management plans was also a key requirement of forest certification. For example, for forests over twenty hectares, both the FSC and PEFC standards required the forest owner to have a long-term SFM forest plan that included the management objectives for the forest, a description of the current state (inventory), and the determination of forest management measures. These certification management plan requirements paralleled the SFA's Green Management Plan ("Green Plan") guidelines and forest goal categories. As a result, many forest owners adapted their Green Plans to follow the certification criteria, and after 1994, management plans took on a new "certification" label. Certification provided structure by augmenting the government's Green Plan guidelines with more detailed instructions regarding long-term ecological landscape-level planning, and by encouraging the appropriate documentation of planned forestry measures and the monitoring of results.

The state further enabled the co-regulation of green management plans when in 2003 the government introduced a new regulation under the Forest Act making it necessary for every forest owner to have at least a simple forest plan (a Forest and Environment Declaration).[35] This encouraged additional private forest owners to certify their forests, as the Forest and Environment Declarations could be easily developed to meet both the regulatory requirement and the certification management plan criteria. Also, because the SFA had no plans to follow up and track the implementation of the Forest and Environment Declaration regulation, certification provided a monitoring and audit enforcement mechanism to reinforce the government's essentially voluntary forest management planning policy.

Overall, the development and implementation of green management plans in Sweden showed how certification was operating as a parallel co-regulatory mechanism to the policy efforts of the Swedish forest authorities, helping to operationalize the forest legislation.

Contributing to SFM Targets: Certification and Voluntary Forest Reserves

Certification not only reinforced the government's broad forest policy goals through green management plans but also played a role in refining and encouraging forest management practices to meet the national forest policy targets. This dynamic was particularly evident with respect to the Swedish national target for forest reserves.

As summarized in Table 6.2, of the total additional 900,000 hectares of high conservation value forestland to be excluded from forest production by the year 2010, the government directed that 400,000 hectares be reserved through legal programs and 500,000 hectares be set aside voluntarily by forest owners. The government hoped that certification would support its voluntary forest reserve target, as both the FSC and PEFC standards required that a minimum of 5 percent of the certified forests be permanently set aside to protect high conservation values. As evaluated below, however, implementation challenges emerged in terms of the alignment of certification and forest policy reserve requirements. Ultimately, certification proved to be a partial mechanism for meeting the government's protected area targets.

Table 6.2

Swedish "Living Forest" protection programs and status

Site protection programs	Protected area target	Program description	2007 status
Legislated	400,000 ha		
Nature reserves	*320,000 ha*	Protect large areas of remaining natural forest.	114,767 ha (36% of target)
Habitat protection areas	*30,000 ha*	Protect smaller, ecologically uniform "islands" of certain biotypes (2-5 ha).	13,500 ha (45% of target)
Nature conservation agreements	*50,000 ha*	Reserve or create sites with HCVs.	17,600 ha (35% of target)
Voluntary set-aside	500,000 ha	Landowner voluntarily sets aside section of forest for benefit of biodiversity or other natural values.	750,000-800,000 ha outside of montane zone safeguarded on voluntary basis, but large uncertainty as to whether HCV forest

Source: Swedish Environmental Protection Agency 2007.

As Table 6.2 shows, the government had difficulty reaching the legal targets, with only 35-45 percent of the reserve areas set aside as of 2007. A major barrier was a lack of sufficient funding to compensate private owners for lost production (Swedish Environmental Protection Agency 2007). In addition, although the voluntary target appeared to have been exceeded, authorities remained uncertain as to whether the forests that were voluntarily set aside by private owners necessarily qualified in terms of representing appropriate ecologically important forests, and also whether the conservation areas were permanently reserved.

A fundamental problem was the lack of an accurate accounting and reporting system for the voluntary set-asides. Certification third-party audits proved insufficient because certification reserves did not necessarily meet the government's voluntary forest reserve objectives. Although the certification forest reserve requirements overlapped with the government's target, there were areas of operational difference between the public and private systems, especially concerning the definition of high conservation value

(HCV) forest, the forest reserve accounting criteria, and the permissible forestry activities on the set-aside areas.[36] As explained by the SFA, different interests put different values on what to set aside.

The distinctions between the public and private approaches to forest reserves created some operational-level confusion, but having the parallel systems also encouraged ongoing improvement of forest management requirements. For example, following recommendations from the FSC and PEFC, the SFA decreed a minimum level of 5 percent reserve per forest estate for both the managed and unmanaged nature conservation goal classes. Driven by certification and then adopted by government, this level became a kind of "political consensus" (Ingemarson 2004,13).

As noted by the SFA, "the systems are building on each other over the long run so we can live with minor operational problems in the short run."[37] The synergistic dynamic of certification and forest policy was also evident in several other instances, particularly concerning the forest structure targets regarding standing trees.

Refining SFM Targets: Certification and Forest Structure Objectives

Certification not only helped to refine forest policy in Sweden but also facilitated continual improvements. For example, in the absence of specific targets, Swedish forest authorities used certification criteria to help support and specify legislated SFM policy requirements. The certification standards included criteria for increased deadwood, conversion of spruce stands, and retention of deciduous trees that supported the SFA's interim forest policy targets. The government also used the certification biotope definition in its financial grant process for legal forest reserves, and certification criteria assisted its chemical inspectors in interpreting the legislation regarding which chemicals should or should not be allowed. Thus, rather than acting independently, the public and private forest governance systems were building on each other. As Hysing and Olsson's study (2005, 522) on biodiversity policy in Sweden concluded, "successful certification is necessary in Sweden for the successful implementation of forest policy ... both are striving to the same targets and to implement the same objectives in collaboration."

Although there has been ongoing certification/policy interaction in many areas related to forest structure (such as deadwood, deciduous trees, spruce forests, and key biotopes), the dynamic was particularly evident with respect to the issue of standing reserve trees. In the late 1980s to early 1990s, the forest legislation had minimum requirements but no clear policy targets for final fellings. In particular, there was no standard for reserve trees – namely, the number of trees that should be left standing per hectare after harvest. The government had tried to consult and determine the level but was told by industrial and family forest owners that the number could not and should not be defined, that every forest was different so the policy should be left as a general statement rather than a specific target.[38] Ten years later, however, the certification process came up with the rule. In 1997, the FSC revisited the issue and was able to reach consensus of ten trees per hectare. As the SFA explained, companies opposed defining numbers when in dialogue with government but acquiesced when the discussion shifted to the certification arena. In 2003, rather than relying completely on the private rule, the government incorporated the certification criteria in its target for high conservation value forest areas.

Thus, over time the government achieved greater policy specificity through interaction with certification. It was a stepwise adaptive process, with policy building on the areas where certification had reached agreement. As the SFA reflected, "the policy process has been that certification defines, then government takes this and develops an even clearer definition. In the future, certification will probably take this and improve it even more."[39] In other words, there was a synergy between the public and private systems that was mutually beneficial.

In summary, certification supplemented rather than supplanted forest policy in Sweden by aligning with the legislative goals and also by going beyond government requirements to specify forest owner obligations and subsequently drive policy change in certain areas. Ultimately, the government's interpretation of what should be considered a "reasonable requirement" of the private landowner was heavily influenced by the certification standards and the debates over establishment of certification criteria.

Supplementing the SFM Discourse

Forest politics in Sweden is described in terms of the "Nordic model" of consensus-based decision making involving close interaction between the regulator and the regulated (Eckerberg 1990). The Swedish Forest Agency relies on consultation and has traditionally had close ties with the forest industry in terms of protecting the national interest in a sustained timber supply, and with the small forest landowners through the local-level delivery of county forestry board informational, technical, and forest management planning services. Given the historical alliances and inclusive approach towards policy between the government and the country's forest landowners, it is not immediately evident that there would be any prospect for forest certification to influence the Swedish forest policy discourse. Accordingly, interviewees had varying perspectives on the implications of forest certification for stakeholder engagement. Some emphasized that certification had negative implications as it disrupted a well-functioning system of division of powers between the government, forest owners, and the public. Others noted that certification reinforced the status quo of traditional alliances instead of giving a voice to new groups. Still others emphasized the positive effects in terms of an increased policy role for social and environmental groups that had been traditionally excluded.

While certification had critical opponents and limitations, overall it enhanced the SFM discourse in Sweden because it provided an alternate political arena for policy engagement.[40] Specifically, the PEFC and FSC created new deliberative forums that enabled a spectrum of stakeholders with varying timber and non-timber forest interests to directly participate and also take on greater sustainable forestry decision-making responsibility. Certification also helped to test SFM assumptions and challenge the broadly conceived national forest vision by forcing debate over new forest rules that tackled head-on the difficult questions concerning forest value trade-offs.

Multi-Stakeholder Forest Policy Engagement

Certification provided a parallel, privately led political arena for reaching forest management agreements in Sweden. Forest certification standard-setting bodies presented an alternate forum where a wide range of forest stakeholders could engage in SFM deliberation beyond traditional government-led commissions and regulatory processes. Whereas industry and family landowner associations had traditionally had privileged access to government, FSC and PEFC membership was open, offering some groups an unprecedented opportunity to contribute to forest policy making. Certification also facilitated a balanced representation of interests. Ecological, social, and economic groups had the opportunity to participate on a level playing field in the certification decision-making processes.[41] As Gulbrandsen (2008b, 114) explains, certification in Sweden has been "a loosely structured system" compared with the traditional hierarchical system of government, which gives privileged access to certain stakeholders. Certification has "less formal or practical barriers for actors seeking to provide input."

For example, through the written constitutions of the certification bodies, certification facilitated increased input to the forest dialogue from labour groups and, in particular, the Sámi indigenous people in northern Sweden. Although the government had initiated a consultation process with the Sámi in the late 1970s under the 1979 Forest Act, the FSC broadened the geographic coverage to include consultations regarding winter grazing land. This brought more Sámi into the process.[42] Also, the FSC offered the Sámi an equal vote and a place at the table within the social chamber – enhancing Sámi cooperative decision-making authority and raising the profile of reindeer herding land access issues.

It is also important to note, however, that "local engagement" has been a point of contention. Although certification was intended to facilitate open participation, among other things, some interviewees expressed concerns that certification had in fact discouraged certain local interests that had historically been taken into account through the close relationship between communities and the regional forestry board and county board authorities.[43]

Overall, these limitations and challenges of certification highlight its role in supplementing rather than replacing traditional state-led processes. Certification added to the state forest consultation processes not only by encouraging additional groups such as labour organizations and the Sámi to join in the forest dialogue, and strengthening their decision-making responsibility and authority, but also by educating and building trust among the country's various forest stakeholders. By providing a forum for ongoing interaction and by placing pressure on the various parties to reach consensus decisions regarding some of Sweden's most difficult forest issues – such as natural old forest, Sámi access, biodiversity, and high conservation value forest reserves – certification helped to expand SFM deliberation and stakeholder learning. This, in turn, assisted the forest authorities in establishing legislated forest policy targets that, in many cases, had already been debated in the private forum. As explained by the SFA, "government consultation on targets improved because stakeholders had already met through certification – certification is helping with the policy process."[44]

Certification also aided state processes by providing a more adaptive policy forum in terms of greater flexibility to reach SFM agreements. Interviewees commented that while the government tried to engage with stakeholders on forest issues, it did not have the mechanisms to "find the edge" where consensus could be reached. This was demonstrated in the case of the reserve trees target (described earlier). Rather than seeking full consensus, the government had to be pragmatic – concerned with finding workable solutions that could be implemented. Unlike certification programs, which were directed to reach consensus within the group, government agencies would consult and then go away and make the decisions themselves. By requiring consensus within an open, balanced forum with equal voting rights, certification gave unprecedented authority and responsibility to Sweden's non-state forest actors to cooperatively engage in a deliberative political process to determine SFM rules. As past LRF chairman and professor Tage Klingberg (2003) explains, certification contributed to a shift in power over forestry decisions. Instead of government-led decisions through consultation with forest owners and industry, certification meant the inclusion of other actors in forest policy development and delivery, such as NGOs (developing and promoting certification), scientists (developing certification criteria), and consultants (verifying forest practices). By requiring the various parties to work collaboratively towards solutions on contentious issues, certification also provided a mechanism for building social capital and fostering trust between the different groups.

Certification continues to be opposed by some for destabilizing the traditional power balance and close alliances between government, industry, and family forestland owners but, overall, it has had a positive influence on forest governance – enhancing the SFM discourse in Sweden by providing an open, parallel forum for forest policy deliberation, and by increasing the authority and responsibility of a wide range of non-state forest actors to reach consensus on sustainable forest management.

Strengthening and Challenging the Forest Vision

The government's goal of balancing environment and production was broadly accepted and articulated in the amendments to the 1994 Forest Act and the 1999 Sustainable Forestry Objectives.[45] The PEFC and FSC standards incorporated this high-level national consensus but also took the next step (prior to the government's efforts in 2005) of specifying particular indicators and targets. Through this process, certification challenged and strengthened the national forest vision.

Despite the macro-level consensus among Sweden's forest stakeholders on the need to balance forest values in the long run, there was disagreement on how to translate the forest ideal into action in the short term – that is, what forest practices should be carried out now to ensure sustainable forestry in the future? Forest actors diverged in their priority weighting of forest values and hence in how they perceived the optimal trade-offs and on-the-ground delivery of the forest vision. Perspectives differed between certification systems as well as within government. As the SFA explained, "the government's vision for the forest is a balance of the forest values but, depending on who you ask, the balancing of the values will change." Although the government projected a common vision, the SFA mainly supported sustained timber production while SEPA was mainly concerned with nature conservation, which is not surprising given their different mandates. Similarly, the PEFC program stressed forest owners' rights whereas the FSC program emphasized the protection of biodiversity and social values. The existence of differing perspectives and interests both between and within the public and private governance systems fostered political contests that tested the feasibility of and commitment to balancing forest values to achieve the long-term forest vision.

In addition, the different views on how to actually deliver Sweden's forest vision were not just political but also rooted in technical uncertainties. For example, while there was scientific agreement with regard to certain SFM practices such as leaving deadwood, setting aside biodiversity hotspots, and restricting forest road construction, the science was still uncertain with regard to other SFM concerns, such as the proportion of forestland to be protected to conserve biodiversity, the use of indicator species, how to identify particularly valuable forest areas for conservation, and how to quantify the necessary protection measures (Gulbrandsen 2008b, 111).

Rather than skirting the technical uncertainties and political difficulties of reconciling divergent sustainable forestry perspectives (as the government had done with the 1994 Forest Act), certification tackled these challenges head-on. By patiently striving for consensus on a set of specific SFM targets that addressed economic, environmental, and social forest values, certification framed the difficult debate over the forest vision in terms of how to actually balance forest values and determine the trade-offs necessary to achieve the forest sustainability goals. As debate continues over the revision and specification of the private rules, certification processes continue to strengthen Sweden's forest vision by encouraging and leading the deliberation and practical testing of an adaptive sustainable forest management ideal.

Summary

Forest certification mechanisms in Sweden have functioned parallel to regulation rather than as a substitute for state forest policy authority. Through interplay between the public and private systems, certification programs have contributed to the ongoing refinement of the national forest goals and objectives. Acknowledging the limitations of certification as a stand-alone forest policy instrument, government authorities in Sweden closely monitored certification developments and responded to the limitations of certification rule making by continually improving upon forest policies and regulations to ensure the implementation of the national forest policy agenda. Overall, the resulting certification/policy dynamic constituted a co-regulatory forest governance system with coincident public and private rule-making authorities.

Although the Swedish government took an accepted position of non-interference, the state played a key role in enabling certification with the introduction of legislation that encouraged private self-regulatory initiative. The forest authorities also kept a watchful eye on certification rule making, revising state forest policy as appropriate to guard their sovereignty and to ensure continued public/private forest policy alignment.

The resulting certification/policy dynamic had several implications for governance, including greater specificity in the country's SFM targets, such as those regarding protected areas and standing reserve trees, and an expanded SFM discourse through a new deliberative forum where stakeholders had increased responsibility for reaching consensus on difficult sustainable forestry questions. Although certification SFM requirements and audit processes did not fully align with or meet all of the government's forest management goals, the private standards played an important role in challenging and strengthening the country's collective forest vision by encouraging ongoing identification and testing of SFM value trade-offs.

Overall, the Swedish case highlights the limitations of certification as a stand-alone policy mechanism but it also demonstrates its significant benefits for state forest governance. By adopting a co-regulatory approach that enabled and leveraged certification alongside traditional state authority, the Swedish government fostered a dynamic of contestation and cooperation that encouraged ongoing continual improvements to both the public and private forest governance systems and the nation's forest management goals.

7
Conclusion

In recent years, corporations and nongovernmental organizations (NGOs) across industry sectors have been cooperating (separate from the state) in the development of a vast array of "non-delegated" corporate social responsibility (CSR) codes and standards. These voluntary private initiatives parallel government-led efforts to promote collaborative environmental governance arrangements such as public/private partnerships and negotiated environmental agreements,

As reviewed in Chapter 2, governance research has focused on why these various state-led and non-state new governance approaches have emerged, how they have been developed, why companies participate, and the comparative effectiveness of voluntary and prescribed rules and standards. Little attention has been paid to the role of government in private environmental governance, however, and to how the public and private rule-making systems are interacting. For the most part, policy analysts have assumed a clear distinction between public and private governance approaches. While this lends theoretical clarity, it has dismissed the empirical reality that the boundaries between public and private governance mechanisms are increasingly blurred, particularly because of co-regulatory governance approaches.

Forest certification is an example of a governance mechanism with overlapping public/private boundaries. For example, the standards incorporate forest law and international forest criteria, and governments are also actively engaging in certification. There has been very little investigation of the public/private policy dynamic, however. In order to address this gap and evaluate how public and private authority interact in the case of forest certification, this book has evaluated how and why governments in the world's leading certified nations (Canada, the United States, and Sweden) have responded to certification, and examined the implications of certification co-regulation. The forest certification governance story has revealed how governments in industrialized countries have strategically engaged in certification alongside prescriptive forest law, resulting in additional forest governance resources and forest policy innovation.

The new governance concept of CSR co-regulation clarified and defined how public and private authority coexist within multicentric co-regulatory governance systems. Three new analytical tools applied the concept of CSR co-regulation to the case study evaluations: a governance typology, a matrix for illustrating a co-regulatory governance system, and a co-regulation spectrum for mapping government CSR response.

In answer to the book's central questions, the empirical evidence demonstrated three main findings. First, governments in Canada, the United States, and Sweden have adopted increasingly direct approaches towards certification, including endorsing certification standards, establishing legislation to enable certification implementation, adopting certification on public land, and mandating certification.

Second, the reasons *why* governments have engaged in certification are similar, but the factors explaining *how* governments chose to respond are different. Governments have participated in certification for a range of reasons, but primarily to ensure policy alignment, minimize potential market distortions, capture potential market reward, and/or sustain public trust. Factors influencing how governments have co-regulated certification are contextual and have included market and non-market drivers such as environmental nongovernmental organization (ENGO) advocacy pressure and government leadership, as well as customer demands, industry expectations, the availability of private funding, and state budgetary pressures.

And third, although certification on its own has clear limitations, certification co-regulation has resulted in a range of governance benefits, including greater efficiencies in state forest administration. In addition, through rule-making competition, co-regulation has encouraged the continual improvement of state forest management practices and policy-making processes, including an enhanced forest management discourse and more adaptive forest policy. The cases also suggest that it is these governance improvements rather than measurable economic gains that have provided governments with a justification for their continued engagement in certification.

Overall, the case study evidence reveals how forest certification is neither a purely non-state nor purely market-driven governance mechanism. Rather, certification is functioning as a co-regulatory forest governance system, engaging both public and private forest authorities.

Returning to the opening chapter, the cases and analysis have helped to shed light on both of the initial puzzles – why certification has been classified as a non-state market-driven mechanism (NSMD) given the overlap with forest laws and government engagement, and what role it is serving in highly regulated northern countries given that it was primarily intended to fill a governance gap in tropical regions. The findings suggest that certification adoption has occurred more in developed rather than developing regions not only because more stringent forest laws lower the marginal costs of certification implementation but also because public sector capacity plays an important role in certification development, implementation, and enforcement. Governments in industrialized countries have supported certification as a complement to rather than a substitute for state forest law. Fundamentally, certification is a co-regulatory governance mechanism, and developing regions have lacked the necessary baseline institutions and/or essential state regulatory resources to achieve successful uptake.

Regarding the second puzzle, as evaluated in Chapter 3, NSMD governance *does* accurately explain the "non-delegated" and global supply chain features of certification, but the theory also falls short. By assuming a zero-sum authority (that is, if governments exert authority, then NSMD no longer exists), it ultimately fails to capture the dynamic of coincident public and private authority within certification co-regulatory systems. By looking beyond the theoretical categorizations of certification as a "non-state market-driven mechanism" or "private hard law" to the empirical reality of government certification engagement and certification/policy interaction, it is clear that governments in the leading certified nations are not in retreat. They are engaging in certification in order to achieve and advance sustainable forest management through co-regulation.

Although context differs, the foregoing lessons and findings have critical application to and implications for achieving optimal governance in other resource sectors where CSR certification standards are also emerging to promote better corporate stewardship.

This final chapter has four sections. In order to provide a comprehensive picture of the certification co-regulatory dynamic, the first brings together the empirical evidence and results from the individual case study evaluations and presents an overall synthesis of the findings. The second section draws on these findings to evaluate the opportunities and challenges of optimal certification co-regulation in accordance with the strengths and weaknesses of certification. The third section provides operational insights to guide policy makers in achieving optimal certification co-regulatory systems. Finally, suggestions for future forest certification and CSR governance research are provided.

Co-Regulating Forest Certification

The Spectrum of Government Engagement

Although the Canadian, US, and Swedish governments communicated their position on certification as one of non-interference, probing further into national and subnational certification responses revealed many examples

Table 7.1

Government forest certification co-regulation strategies

Certification aspect	Indirect co-regulation	Direct co-regulation
Development (rule making)	• Attend standards development/revision meetings • Provide guidance as requested • Promote forest policy alignment	• Provide a legal framework • Provide resources • Participate as voting member in standards development/revision
Implementation	• Provide information and training	• Remove legislative barriers • Establish cooperative agreements • Provide certification incentives • Certify public forestland • Incorporate private rules into policy
Enforcement	• Assist companies in preparation of certification audit evidence • Clarify forest policy during certification audits • Threaten to mandate certification	• Provide accreditation services • Mandate certification • Incorporate certification audit in legislative compliance audit • Develop public procurement policy for certified forest products

of government engagement with certification. Responses were similar in terms of indirect baseline cooperation but varied in terms of direct co-regulatory approaches. Table 7.1 provides a summary of the range of government forest certification co-regulation strategies employed, from indirect to direct approaches at the development, implementation, and enforcement stages.

Development

At the development stage, all governments placed themselves initially, at a minimum, in an indirect facilitating role by providing guidance to ensure certification credibility and policy alignment. As well, by default, all played a direct role in standards development by providing a supporting legal framework to enable contracts and establish a baseline of sustainable forestry legal compliance. Beyond this, some governments provided financial, technical, and/or human resources to support the development of the standards, and also participated directly on the standards development committees. For example, the Swedish Environmental Protection Agency volunteered its director's time to chair the national FSC working group. A government representative from a US state, Oregon, was a member of the Sustainable Forestry Initiative (SFI) board; Canadian provincial forest ministry representatives participated in the Canadian Standards Association (CSA) Technical Committee and attended Forest Stewardship Council (FSC) meetings as observers; and representatives from the Swedish Forest Agency (SFA) participated in subcommittees of the national FSC working group. The Canadian federal government also endorsed certification in its National Forest Strategy and provided support for the development of the national CSA standard through a public agency, the Standards Council of Canada (SCC). All cases showed that governments were contributors to, and in some instances drivers of, certification standards development.

Implementation

At the implementation stage, governments indirectly facilitated certification adoption by providing information, training, and/or technical assistance to forest owners and operators. More direct approaches included establishing partnership agreements with the certification bodies (Ontario), providing financial incentives (Wisconsin managed forest law for small private forest operators), removing legislative barriers (Sweden's 1994 Forest Act; British Columbia's 2005 Forest and Range Act), and adopting certification on state-owned forestland (the certification of Canadian provincial Crown land, US state-owned forestland, and public land managed by Sveaskog in Sweden).

Enforcement

Finally, at the enforcement stage, governments indirectly supported certification by assisting in the preparation of audit evidence, offering technical support during certification audits, and ensuring the alignment of forest certification audits with state regulatory compliance audits. In certain cases, governments used the indirect approach of a regulatory threat as a means of encouraging certification adoption (Sweden, Maine). In terms of direct engagement in certification enforcement, several governments mandated certification (Ontario, New Brunswick, and Michigan), provided auditor accreditation services (SWEDAC and the SCC), piloted the incorporation of certification as a component of the legislative compliance audit (Ontario, BC, and New Brunswick), and established public procurement policies to promote demand for certified forest products (Vermont, Michigan, and Washington).

At a minimum, state authorities closely observed the development and implementation of certification standards, and the lead regulatory forest

Figure 7.1

Summary of government responses to forest certification

	Indirect → Direct				
	Observe	*Cooperate*	*Enable*	*Endorse*	*Mandate*
Government role	Monitoring certification to ensure policy alignment and prevent market discrimination	Providing information and assistance as requested	Supporting certification in the forest policy mix	Certifying state-owned forests	Legislating certification on public land
United States	✓	✓		State governments certifying state-owned forests	
Canada	✓	✓	A range of provincial government response to certification, including Ontario and New Brunswick mandating certification on Crown forestland		
Sweden	✓	✓	Forestry frame law encouraging private forest governance initiative		

agencies in each country (Canadian provincial forest ministries, US state-level departments of natural resources, and the Swedish Forest Agency) cooperated with certification organizations by providing information and technical advice. Beyond a baseline level of cooperation, government engagement over the period 1993-2008 included various increasingly direct co-regulatory roles in terms of enabling, endorsing, and even mandating certification. In particular, as shown in Figure 7.1, the most prominent examples of *direct* co-regulation in each jurisdiction included enabling certification through results-based forest legislation (Sweden), endorsing certification by certifying state-owned forests (twelve US states), and mandating certification on public forestland (Ontario, New Brunswick, and Michigan). As summarized in the next section, although the underlying rationale for cooperating in certification was similar among governments, the drivers explaining why governments adopted more direct co-regulatory approaches included a range of factors that played out differently within each jurisdiction.

Figure 7.2

Government rationale for engagement in certification

Manage potential risks
- Ensure policy alignment
- Prevent market discrimination or trade distortions

Improve industry competitiveness
- Sustain market access
- Facilitate potential market advantages and gains

Demonstrate sustainable forestry
- Demonstrate and improve state forest policies and practices
- Establish state forest management leadership

Increase public trust
- Increase citizen understanding and engagement in forest decisions
- Build credibility with ENGOs

The Rationale and Drivers of Certification Co-Regulation

Rationale

As global timber producers with well-established forest laws, facing similar global economic pressures and increasing societal expectations regarding the protection of environmental forest values, Canada, the US, and Sweden had similar rationales for at least observing and cooperating in certification (see Figure 7.2). All wanted to ensure policy alignment, protect forest owners from potential market discrimination, and prevent domestic forest producers from suffering from any trade distortions. In other words, it was important to manage for any political and/or economic risk. These governments also supported certification in order to capture any potential economic gains. As a market-based instrument, certification was a means of gaining a possible advantage for domestic forest producers that would enhance their long-term global competitiveness. At a minimum, therefore, the governments wanted to acknowledge and step out of the way of forest owner certification efforts.

Beyond economic considerations, the common rationale for engaging more directly in certification included additional environmental and sociopolitical justifications, such as demonstrating and improving state forest management policies and practices, increasing citizen understanding and engagement in sustainable forest management decisions, and building credibility and trust with the public and environmental advocacy groups.

Table 7.2

Certification co-regulation: regional considerations and drivers

Country	Co-regulatory approach	Key consideration	Drivers
Canada	A range of provincial government approaches, including mandating certification	Guard policy sovereignty and ensure regulatory alignment	Industry expectations International ENGO pressure Policy alignment according to stage of the policy cycle
USA	Adopt certification on state-owned forestland	Spur state forest economy and set leadership example for family forest owners	Issue-based: • Customer pressure • Interstate competition • Ailing state economies • ENGO advocacy Opportunity-based: • Private foundation funding • Market access and potential premiums • Government leadership
Sweden	Enable certification through frame-law forest legislation	Support self-regulatory achievement of national forest objectives	Alignment with Forest Act Industry support ENGO trust Sveaskog certification leadership

Certification Co-Regulation Considerations and Drivers

Although the governments in the three countries shared similar rationales for cooperating with certification, they directly co-regulated certification in different ways based on specific local forest regime considerations. For example, as summarized in Table 7.2, in Canada, where provincial governments are both the principal landowners and regulators, a fundamental consideration in certification co-regulation was to guard the policy agenda and protect policy sovereignty, and thus ensure that certification standards were developed and delivered so as to align with forest regulation. In Sweden, where the state was taking a "softer" approach of informing and steering rather than coercing private forest owners, the government responded to certification so as to maintain its enabling policy role. And in the US, where state governments are the principal forest regulators of private forestland, an important consideration in adopting certification was to demonstrate state leadership to encourage private family forest owners, a sizable landowner group across the country that has historically been difficult to regulate and persuade to become actively engaged in forest management.

Beyond the rationale and fundamental considerations for certification engagement, the case studies also showed that various factors interacted with each other and with the background social, political, economic, and environmental conditions within each jurisdiction to influence governments' direct co-regulatory response. As Table 7.2 also shows, in the US a range of issue-based and opportunity-based drivers influenced state governments to certify their forests, including customer pressures, interstate competition, and ailing state economies, as well as the availability of private foundation funding, concerns over maintaining market access, and hopes of gaining price premiums. Among the Canadian provinces, three significant factors prompted varying co-regulatory government responses, including industry expectations of government cooperation, strong international environmental advocacy pressures, and sovereignty concerns over protecting the domestic policy agenda. In Sweden, the government's co-regulatory role in enabling certification was shaped to a large extent by the overall shift in the forest regime with the introduction of the 1994 Forest Act. The Swedish government also felt assured in enabling certification as it had established trust with domestic environmental groups during the forest campaigns in the late 1980s, and also felt a certain level of confidence in the domestic industry as companies were demonstrating strong leadership and commitment to supporting FSC certification (including the state-owned forest company, Sveaskog).

In summary, similar macro-level conditions in the US, Canadian, and Swedish forest political economies prompted governments in these regions to take notice and at a minimum cooperate with certification efforts in order to manage for any potential political and/or economic risks. Institutional differences in the local forest regimes, such as the balance of public/private forest ownership, the level of forest administrative authority, and the historical policy style, helped to explain *why* each government adopted its particular co-regulatory approach. Additional context-specific political, economic, social, and environmental drivers also played a role in determining *how* governments directly engaged in certification.

Governance Implications of Certification Co-Regulation

Chapters 4 to 6 demonstrated how all governments in the leading certified forest nations co-regulated certification. This was the case even though all of them characterized certification as a market-based instrument and communicated their role as one of non-interference. By evaluating the co-regulatory dynamic between the private and public systems, the cases revealed a competitive yet synergistic interplay between the public and private rules and processes. In other words, the two systems operated in parallel and behaved as complements with mutual dependencies and mutual benefits. Fundamentally, certification systems have relied on a legal framework and have drawn on government acceptance and support for legitimacy. Governments have also benefited from certification in terms of improvements to state forest administration, continual improvements in state forest practices, enhancements to the forest policy discourse, and, overall, greater governance capability and effectiveness through the incorporation of more adaptive and responsive forest policy rules within the policy mix.

Forest Administration Benefits

Adoption of certification on state-managed public forestland improved state forest administration. In particular, meeting certification requirements and undergoing third-party audits encouraged continual improvements to state forest management planning and forestry practices. For example, as shown across the twelve US states that certified their forestland during the period 1993-2008, certification served as a springboard for achieving positive governance outcomes such as improvement in staff morale, better departmental coordination, greater access to state funds, shorter forest planning time frames, more up-to-date technical resources, greater public engagement, more regulatory forest tracking and reporting, improved forestry practices, and demonstrated state government leadership. In particular, meeting the certification public engagement and reporting requirements increased the transparency and accountability of the state's forest management processes. Implementing the required improvements to the tracking and monitoring systems, forest models, inventories, and forest plans, and responding to the regular independent third-party certification auditor feedback all contributed to improving the state forest agencies' sustainable forest management practices. Overall, by adopting certification on state-managed public land, governments realized benefits largely associated with having an effective environmental management system (better tracking, coordination, delivery, and communication of programs, as well as continual forest management performance improvements).

Enhanced Polity, Politics, and Policy

In terms of the three key aspects of governance – polity, politics, and policy – the cases also showed that the incorporation of certification as an instrument in the state forest policy mix enabled greater private decision-making authority, encouraged increased forest stakeholder participation and deliberation, and facilitated continual improvement in forest policy rules and process (Table 7.3).

Table 7.3

Certification co-regulation governance outcomes

Governance aspect	Outcome of certification
Polity (decision-making forum)	Expanded political arena beyond traditional state-led decision making, to include private rule-making authority
Politics (decision-making process)	Expanded multi-stakeholder policy deliberation and responsibility, increasing stakeholder knowledge and building social capital
Policy (forest decisions)	Supplemental resources and more adaptive forest regulation

Polity. Certification has provided an additional political forum for the deliberation of sustainable forest management issues and the establishment of forest management rules. Led by private actors such as forest companies, private landowner associations, and/or environmental organizations, the private certification arena has extended forest policy making beyond the traditional sphere of state-led agenda setting and rule creation to encompass private forest governance authority. Although "non-delegated" and distinct from state policy making, certification rule making has not been detached from the state. Certification systems have been operating in parallel with state processes and have gained legitimacy not only through market supply chains and the acceptance of corporate actors and NGOs but also through direct government engagement. By co-regulating certification, governments in Canada, Sweden, and the United States have all played a role in legitimizing private forest governance authority.

Politics. Certification authority has also been legitimized through a virtuous cycle of enhanced stakeholder engagement, learning, trust, and collective sustainable forest management decision making (see Figure 7.3).[1]

Certification bodies mimic democratic institutions as they operate under constitutions that encourage multi-stakeholder input and establish fair and equitable decision-making procedures and transparency requirements. As the cases illustrated, however, unlike traditional state policy making, the certification standard-setting process has not been driven from the top down by a single state authority but rather through market and NGO-led private authority. Governments have contributed to but have not formally delegated authority to nor dominated certification decision making. As a consequence, certification processes have had to gain their legitimacy and rule-making

Figure 7.3

The politics of certification authority: a virtuous cycle

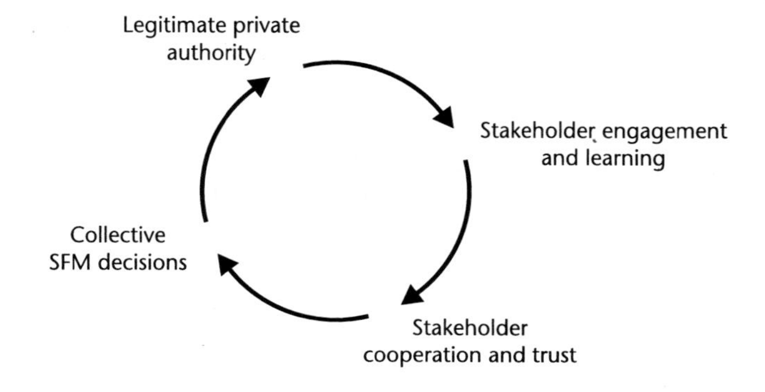

authority through ongoing market and social acceptance. Ensuring open access for multi-stakeholder engagement has therefore been a priority.

Groups have been drawn to actively participate in the certification process not just because it has been accessible but also because of the opportunity to establish rather than just influence forest rules. As described in the cases, for some stakeholders, particularly social groups often excluded from the formal state forest dialogue, the regional certification arena offered unprecedented forest decision-making access and responsibility. For example, in all countries, the three chambers of the FSC consensus-based national and regional standard-setting bodies provided economic, environmental, and social stakeholders with an opportunity to participate on an equal footing in forest debates and decisions that had in many cases been the traditional domain of close industry/government alliances. In Canada, the CSA local multi-stakeholder SFM advisory groups supplemented the provincially legislated forest licensee stakeholder consultation requirements, enhancing local stakeholder engagement and influence. And in the US, the regional SFI implementation committees (SICs) brought local forest stakeholders together to develop SFM programs, facilitate SFI adoption, and encourage continual SFM improvements beyond state requirements.

The increased engagement and responsibility of stakeholders in the certification process has, in turn, enhanced the forest dialogue and encouraged stakeholder learning. Certification has provided a deliberative forum for stakeholders to voice, debate, and better understand the range of perspectives on SFM issues. In developing and revising the certification standards, groups have met regularly to deliberate and reach collective decisions that they know they will be responsible for implementing. The ongoing interaction and shared accountability has encouraged stakeholder cooperation and trust. The cases showed that, spurred by the collective desire of stakeholders to reach agreements themselves rather than having governments impose requirements through top-down regulatory intervention, SFM decisions were reached within privately led certification bodies that governments had previously been unable to achieve.

Fundamentally, the increased private forest governance responsibility and authority facilitated a different style of interaction among stakeholders than within the traditional state-led processes. Returning to Figure 7.3, deliberation within the certification arena has ultimately established a virtuous cycle wherein private rule-making authority has enhanced forest stakeholders' social capital, enabling difficult SFM decisions to be reached and ultimately reinforcing the continued acceptance and legitimacy of certification rule-making authority.

Policy. Certification has also aided governments in formulating, implementing, and enforcing forest policy. As summarized in Table 7.4, certification deliberation, certification rules, and certification audits have challenged and improved the policy process. Certification deliberation increased stakeholder knowledge, cooperation, and trust, and enhanced governments' understanding of the issues and stakeholder perspectives. This reduced educational demands on government, facilitated meaningful stakeholder policy input, and improved government receptivity, all of which ultimately enhanced policy formulation and implementation processes.

Certification rules have assisted in the formulation and implementation of policy by facilitating a competitive, stepwise public/private interplay that has encouraged more adaptive forest governance – that is, rule making that is constantly testing, receiving feedback, and making adjustments, and therefore continually improving. For example, as shown in the Swedish case, certification standards went beyond state legislation and state forest policy, which in turn improved the private rules. There has been a dynamic synergy, with each system challenging and advancing the other.

Finally, certification audits have augmented forest policy implementation and enforcement by reinforcing regulatory compliance, encouraging beyond-compliance forest practices, and enhancing the transparency and accountability of forest operations. In particular, certification audits have provided governments with additional information about forestry performance and specific areas requiring corrective action, which has assisted forest authorities in identifying problem areas and in formulating and delivering more effectively targeted forest policies and programs.

In summary, forest certification has *not* functioned as a purely private non-state market-driven governance mechanism. Instead, the certification

Table 7.4

Contribution of certification to forest policy process

Characteristic	Benefit	Policy formulation	Policy implementation	Policy enforcement
Certification deliberation	• Increased stakeholder knowledge	✓	✓	
	• Increased stakeholder cooperation and trust	✓	✓	
	• Enhanced understanding of issues and stakeholder perspectives	✓	✓	
Certification rules	• Interplay with state forest policies	✓	✓	
Certification audits	• Reinforced compliance		✓	✓
	• Beyond-compliance continual improvements	✓	✓	✓
	• Enhanced transparency	✓	✓	✓
	• Supplemental accountability		✓	✓

standards are inherently hybridized (the private rules incorporate public law) and governments have been directly engaged. The co-regulatory dynamic has facilitated positive governance outcomes such as enhancements in state forest administration, more progressive forest discourse, and continual improvements in state forest policies and processes. Despite the forest governance opportunities offered by certification, however, governments have also recognized its limitations and challenges.

The Co-Regulatory Opportunities and Challenges

As outlined in Chapter 2, the essence of the co-regulatory governance challenge is to achieve a balance between dynamic, innovative private rule

Table 7.5

Opportunities and challenges of certification co-regulation

Governance area	Opportunities	Challenges
Policy design	Certification and regulation have complementary policy features.	Certification is not a comprehensive, stand-alone forest policy instrument.
Policy target	Certification is targeting leading and compliant forest operators, freeing public resources to focus on non-compliant forest actors.	Certification is not targeting laggards. Unresolved debate over certification standards makes policy target unstable.
Institutional durability	Certification can adapt to shifting stakeholder expectations.	Certification is losing flexibility as standards converge and partnerships deepen.

making and stabilizing, accountable public regulation so as to maximize the benefits while minimizing the drawbacks inherent in each system. As shown in the cases, governments have recognized both the strengths of certification and its limitations as a stand-alone policy mechanism. They have therefore approached certification rules and decision-making processes as a potential addition rather than as a substitute for state forest laws and policy making. Governments have not retreated from law making but have directly endorsed, adopted, and/or enabled certification, leveraging private resources while at the same time protecting their forest policy-making sovereignty. The key challenges and opportunities of certification co-regulation with respect to policy design, policy target setting, and institutional durability are summarized in Table 7.5 and discussed below.

Policy Design

Certification and regulation have distinct yet synergistic governance features. Areas of regulatory weakness are certification strengths, and vice versa. As presented in Table 7.6, assigning approximate, unweighted values relative to policy criteria (High = 3, Medium = 2, Low = 1) shows the areas of contribution and highlights the corresponding synergies between certification and regulation as potentially complementary policy instruments within a co-regulatory forest governance system.

As reviewed in Chapter 2, regulation is an attractive policy instrument for governments because in strong democratic states it is legitimate, accountable, and enforceable. It also tends to be slow and costly, however. CSR

Table 7.6

Certification and regulation: complementary governance attributes

	Selection criteria	Certification	Regulation
Policy instrument	Legitimate	L (1)	H (3)
	Accountable	L (1)	H (3)
	Efficient (timing)	H (3)	L (1)
	Cost-effective (expense)	H (3)	L (1)
	Output effectiveness (uptake)	L (1)	H (3)
	Outcome effectiveness (results)	M (2)	H (3)
		11	14
SFM policy tool	Adaptive rule making	H (3)	L (1)
	Local decision making	H (3)	L (1)
	Integrated forest management	H (3)	M (2)
	Comprehensive SFM	L (1)	H (3)
		10	7
		21	21

Note: The numbers represent approximate, unweighted values assigned to the policy criteria: High (H): 3; Medium (M): 2; and Low (L): 1.

certification standards are fundamentally appealing because they are faster to implement and less expensive than regulation, but privately led certification by NGOs and corporations is voluntary and so implementation is uncertain. In addition, certification legitimacy is unstable and accountability is limited (certification bodies are unelected, and market, government, and/or social acceptance are not guaranteed). In terms of outcome effectiveness, both forest certification and regulation contribute to improvements in forestry practices, with regulation being perhaps stronger since it is enforceable and thus ultimately more predictable. Assessing the outcome effectiveness of certification versus traditional regulatory approaches is an important area for future research.

Combining the fundamental policy attributes (from the top half of Table 7.6), regulation appears to be a slightly stronger policy tool than certification. Forest sustainability, however, is a unique governance challenge that has required the consideration of additional policy attributes (bottom half of Table 7.6). Certification aligns more closely with these criteria, thus offering governments a governance opportunity.

Forests are a complex resource to govern, not just because they are both a public and private good but also because the management of sustainable forests is subject to natural disturbance; political, scientific, and technical uncertainties; shifting societal and community-based forest values; and dynamic local forest conditions. SFM is both a process and a moving target.

Depending on the forest community, the sustainability equation will vary. Some policy criteria that are normally desirable may not necessarily be optimal when it comes to addressing SFM issues. For example, while the US case study showed that certification can facilitate the efficient delivery of state policies and programs, state governments also explained that it may not always be appropriate to speed up the public process. Forest planning (particularly on public forestland) often requires lengthy deliberation and coordinated processes, so certification efficiency could be a policy drawback as it may encourage overly hasty decisions about forests.

Thus, the additional criteria that governments have been considering when designing their forest policy mix have included adaptive rule making, local decision making, integrated forest management, and a comprehensive scope. In other words, optimal forest policy approaches strive to be flexible to meet local conditions; promote local community engagement; integrate and balance economic, environmental, and social forest values; and ultimately make a positive contribution to sustaining forest health and productivity.

Returning to Table 7.6, one can see that public forest regulation scores slightly lower than certification with respect to these specific (unweighted) forest policy criteria. This is because legislated rules are typically more stable than flexible, are centrally rather than locally driven, and are traditionally developed and enforced as a series of discrete economic and environmental forest laws rather than integrated SFM requirements. In contrast, certification systems establish adaptive forest rules and processes, encourage local forest stakeholder engagement and responsive decision- making authority, and promote the integration of economic, social, and environmental forestry requirements.

Comparing the rough total scores in the bottom half of Table 7.6, it would appear that certification is a slightly more appropriate SFM policy instrument than regulation. Governments have recognized, however, that certification has a fundamental limitation – its scope. Fundamentally, certification is not a measure or standard of forest sustainability. It does not provide a general indicator of the state of the forest or a guarantee of future forest health. Unlike state monitoring and regulatory compliance audits, certification does not measure or assess on-the-ground forest changes resulting from forest practices. As explained by state authorities, certification is a test for independently verifying comprehensive forestry planning and responsible harvesting practices within a defined forest stand, but there are sustainability considerations that fall beyond certification at the forest landscape level (for example, biodiversity). Thus, while certification can expand on traditional regulatory approaches, governments have recognized that certification is not a substitute for forest regulation.

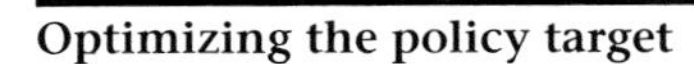

Figure 7.4

Optimizing the policy target

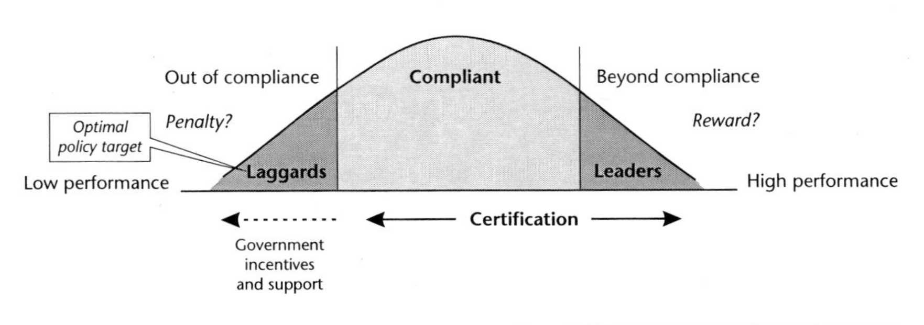

Forest certification has inherent limitations as a stand-alone forest policy instrument, not just because certification is voluntary, leans on a legal framework, and has limited accountability to the wider electorate and a potentially unstable political legitimacy, but also because certification has a partial SFM mandate that falls short of a government's overall stewardship responsibility to protect the forest public goods. As outlined and argued throughout this book, however, certification has significant strengths that can be complementary to rather than a substitute for traditional regulatory approaches within the forest policy mix. As explained next, a further area of co-regulatory challenge as well as a potential governance opportunity concerns the forest policy target.

Policy Targets

An optimal forest policy mix considers not only the range of criteria outlined above but also the policy target. Is the intention to reward the leaders who are going beyond legal compliance? Penalize the laggards who are out of compliance? Or perhaps encourage the compliant majority to sustain and/or improve their performance? Targeting the laggards will typically generate the greatest marginal benefit as there is the potential for the greatest outcome gains.[2] So, what has been the certification target? Has certification served as a "gold standard" to encourage forest leadership, as a market penalty for laggards, or as a benchmark for reinforcing legal compliance? As shown in Figure 7.4, in the highly regulated regions where uptake has been largely occurring, the cases showed that certification has been serving as a progressive baseline standard, reinforcing legal compliance as well as encouraging beyond-compliance forest management leadership.[3]

By targeting the larger segment of compliant and beyond-compliant forest actors, certification has been serving to reinforce forest laws. This reduces some of the regulatory pressure on government, thus giving the state greater opportunity to direct its resources towards problem areas (non-compliance) rather than simply overseeing the compliant majority.

So far, certification has *not* been a major inducement to the lagging performers. Certification requirements have been sufficiently high to constitute a cost barrier, and certification markets have not materialized to the point where there is a sufficient penalty or incentive to ensure participation. In many instances, the lagging forest actors have been small private forest owners that typically lack the technical and financial resources to carry out formal forest management planning. Although they produce a large percentage of the US and Swedish fibre supply, family forest owners in both countries have been a difficult group to regulate. Thus, governments have begun to introduce certification financial incentives such as land tax reductions, initiate targeted certification education and training programs, and support efforts to lower certification costs (for example, through group certification options) in an effort to bring on-board the lagging small private owners through "softer" means than coercive regulation. As illustrated in Figure 7.4, these government initiatives constitute an effort to stretch the certification policy target. Tracing and assessing this growing trend of governments playing a direct role in certification by making it more accessible to the poorer-performing forest owners and operators is an emerging area for future investigation.

A final point of consideration is that while certification has been operating as a progressive baseline standard, its future policy target is a point of ongoing debate. Certification organizations are undecided on the normative question of how certification rules *should* evolve. Some members want to see the bar raised whereas others would like to see a minimum standard maintained. A gold standard would improve performance but could limit participation. A baseline standard would encourage wider adoption among the lagging performers but might dilute the value and market advantage of certification. The discussion raises pragmatic questions about the role of certification as a global forest governance standard, and highlights fundamentally divergent perspectives about the intended purpose of certification. Is certification a forest protection or a forest production standard? Is the goal to conserve high conservation value forests or to create a market advantage for certified forest products from sustainably managed forests? Is it realistic to presume that certification can achieve both?

Certification rules are flexible, which is a positive attribute in terms of enabling local-level responsiveness, but as the rules evolve to set either a higher or lower performance bar, the certification policy target will also shift, thus altering the forest policy mix. The unresolved normative debate about certification's long-term purpose and how certification rules should evolve introduces instability in the certification policy target and ultimately uncertainties into the future design and durability of certification co-regulatory systems.

Institutional Durability

Can governments rely on certification over the long term as a durable forest governance mechanism, or is certification perhaps just a temporary regulatory trend? On the one hand, forest certification systems appear durable as they have achieved institutional capacity (legitimate private rule-making authority) by continually adapting to shifting stakeholder expectations and competitive pressures. On the other hand, as certification membership, rule-making processes, and private forest rules become increasingly entrenched, certification systems are losing some of the adaptive flexibility that contributes to their ongoing acceptance and rule-making authority.

As explained in Chapter 3, because certification is a non-delegated voluntary policy mechanism, certification programs have had to gain rule-making authority through acceptance rather than formal state delegation. Certification rules and processes have therefore needed to be responsive in order to achieve and maintain political legitimacy. Certification organizations have struggled to develop, implement, and revise the standards in an optimal way to encourage SFM results as well as to satisfy the broadest constituency – too rigorous or too lenient a standard and stakeholders might become disenfranchised, breaking the virtuous cycle that reinforces private rule-making authority (see Figure 7.3).

Certification organizations have also had to compete with other certification programs for acceptance. As argued in Chapter 3, while many aspects of the design and SFM indicators within the Programme for the Endorsement of Forest Certification (PEFC) and FSC standards have been converging, philosophical differences and political battles for legitimacy continue. PEFC supporters generally believe that increasing harmonization of the standards has been a source of continual improvement, whereas FSC advocates complain that the competition has resulted in compromise. ENGOs are voicing concerns that the certification requirements are becoming too lax, and are also recognizing that as the systems mature and partnerships deepen, certification is losing some of its leverage as an effective advocacy tool for holding corporations accountable and encouraging continual forestry improvement. The viability of certification programs is increasingly dependent on sustaining rather than critically scrutinizing the acceptability of corporate membership, thus reducing the role of activism and ultimately restricting rule-making flexibility.

As a consequence, there have been warning signs of a waning of ENGO acceptance of certification, and hence mounting concerns about certification durability. For example, in March 2008, the Swedish Society for Nature Conservation (SSNC), the leading environmental group in Sweden, withdrew from the Swedish FSC board, stating that the standard was now too weak and was not being properly enforced. In September 2008, one of the leading global environmental organizations that founded the FSC, Friends of the Earth, withdrew from FSC International over concerns that the standard was supporting unsustainable forest practices by certifying plantations and permitting the harvesting of primary old-growth forests.

So far, despite increasing concerns, certification adoption continues to grow in new forest regions around the globe.[4] Certification programs have demonstrated remarkable resiliency by responding to the negative feedback and adapting to shifting stakeholder expectations.[5] As explained in Chapter 3, certification appears to be moving beyond a pragmatic calculation of costs and benefits towards gaining moral and cognitive acceptance as simply "the right thing to do." The erosion of rule-making flexibility ultimately threatens certification durability, however, and hence co-regulation stability.

In summary, certification's shifting rules, unstable policy target, and uncertain institutional durability present ongoing co-regulatory challenges. Fundamentally, certification co-regulation is not a stable policy arrangement but rather a dynamic governance system subject to constant adjustment and realignment as private and public standards evolve. Evaluating how certification co-regulation matures in developed regions and emerges in developing regions is an important topic for future investigation. Before turning to these topics, however, I examine the pragmatic aspects of certification co-regulation, including a set of practical recommendations for policy decision makers seeking to achieve more effective forest governance solutions.

Certification Co-Regulation in Practice

Tracing the evolution of government response to forest certification in the world's leading certified countries over the past fifteen years has revealed a trend from indirect facilitation to direct co-regulatory engagement. Governments are increasingly enabling, endorsing, and even mandating forest certification, influenced by a range of interacting political, economic, and social drivers. The state is not in retreat. Certification has not been a substitute for forest law. Rather, governments in the case study regions have been co-regulating corporate social responsibility. This marks the emergence of a new political arrangement, one in which public and private rule-making authority coexist in an expanded multicentric political arena.

Achieving an optimal public/private balance within this expanded political forum is a dynamic and tricky process. The challenge for governments is to enable and endorse rather than capture private rule-making authority, while at the same time maintaining state policy sovereignty. The cases in this book have demonstrated how governments have strategically engaged with certification, ultimately striving for a synergy of penalty and reward, and a balance between regulatory freedom and constraint so as to encourage CSR initiative.

Stepping back, enabling CSR mechanisms such as forest certification can be beneficial not only because it leverages private resources but also because the standards are innovative, go beyond the constraints of governments, and involve direct engagement between those that are contributing to environmental problems and those that are directly impacted, such as corporations and civil society. CSR implementation is unpredictable, however; legitimacy is unstable and durability is uncertain. The public sector role in CSR therefore involves fostering the strengths while mitigating the weaknesses of the voluntary mechanisms. The challenge is to design regulatory institutions so as to "protect us from knaves while leaving space for the nurturing of civic virtue" (Ayres and Braithwaite 1992, 53). Ultimately, CSR co-regulation is about fostering corporate responsibility while establishing a solid regulatory backstop to ensure corporate accountability.

The case studies reveal a range of CSR co-regulation options and considerations. The following insights are gleaned from the forest certification case examples and are intended to offer guidance to policy makers in their certification co-regulation decisions:

- Certification is a supplement to, not a substitute for, public regulation. For example, forest certification does not ensure the protection of collective forest benefits such as soil stability, water purification, climate control, wildlife habitat, and species biodiversity. Forest certification can encourage compliance and environmental forest stewardship improvements within designated forest areas, but falls short as a stand-alone policy tool to protect public goods that occur across the larger forested landscape.
- The business case for certification co-regulation is risk mitigation and governance improvement rather than measurable economic gain. So far, forest certification is not generating a significant price premium or increase in market share for the majority of certified forest producers, but rather is serving as a means of sustaining market access and maintaining public trust. In industrialized regions, most forest operators now view certification as an "accepted way of doing business" rather than a way of gaining market advantage. Incorporating certification alongside forest regulation in the policy mix offers potential governance benefits, such as greater stakeholder engagement, enhanced forest discourse, continual improvements to forest administration, and more adaptive forest policy making.
- Adopting an inclusive approach that recognizes all credible certification systems (as opposed to endorsing one system over another) considers the interests of the broadest constituency and prevents market discrimination.

Although the FSC and PEFC programs have increasingly similar design and content and both are accepted in the global market, the standards still have respective strengths and weaknesses and fundamental differences that continue to be debated by the different "camps" (the FSC is promoted by environmental advocacy organizations, whereas large industrial companies and small private forest owners generally support the PEFC standard). Recognizing both programs provides certification options for the range of forest owners and operators who are striving to meet their various domestic and global forest customer demands.

- Leading compliant and beyond-compliant corporate actors have been the major certification adopters. Forest certification has not provided a major SFM inducement to lagging industrial forest operators. State measures are required to address this group.
- Small operators are at a certification cost disadvantage relative to large industrial operators. Family forest owner participation in certification will be minimal unless these owners cooperate in a group certification effort (to achieve economies of scale), are assisted by larger forest companies, or a significant price premium develops for certified forest products. Governments can help offset the inequity by establishing direct financial incentives and/or providing technical support with forest plans and inventories. They can also help facilitate the development of certification program options targeted at this group.
- Independent third-party certification audits can enhance but cannot replace a government compliance audit or state monitoring program. Unlike a compliance audit that documents the effectiveness of forest activities and extent of forest change, a forest certification audit checks on whether a forest operator is following through on various forest management planning and operating commitments in accordance with criteria identified in the certification standard. The certification audit does not measure or evaluate the on-the-ground impact of forest practices or policies, or assess the ongoing state of the forest landscape. Certification audits also do not have the same level of transparency and documentation as government forest audits. They do, however, include additional criteria that go beyond forest law. Thus, integrating certification audits into a compliance program can expand state forest monitoring and enforcement programs and help to demonstrate sustainable forest management beyond legal compliance.

It is important to emphasize that these operational insights regarding CSR co-regulation reflect the case of forest certification and the experience of governments in developed countries with well-established legal systems and well-developed and enforced forest laws. The governance potential of CSR standards and forest certification in developing countries that have limited public capacity as well as very different sustainability issues presents another set of co-regulatory challenges. A critical lesson that *is* transferable to developing countries is that the role of public sector capacity cannot be ignored in understanding why certification adoption has been significantly lagging in these regions.

Future Research

This book is one of the first contributions to a much larger emerging field of research concerning CSR co-regulation and therefore highlights many important new areas and questions for future investigation. In terms of forest certification governance research, there is a large knowledge gap concerning the challenges and prospects for forest certification in developing countries that face higher implementation barriers and lack sufficient institutional co-regulation capacity. Beyond this, three major knowledge gaps concern certification effectiveness, evolution, and durability. There is a need for effectiveness studies that evaluate the direct on-the-ground forest outcomes of certification. Research efforts to compare such outcomes with the impacts of traditional command-and-control regulation would also be very beneficial. There is also a need for studies that examine the indirect outcomes of certification. For example, although forest certification may not subvert existing forest laws, it is unclear whether certification co-regulation might discourage the introduction of new prescriptive legislation that could improve forest practices. Gathering conclusive evidence on forest impacts and indirect governance effects may be difficult, but considerations of effectiveness are essential to the future design of optimal co-regulatory forest governance systems.

Research that traces the evolution of certification rules and government certification engagement is also needed to further develop the concept of CSR co-regulation. For example, studies could evaluate how and why a government's participation shifts and how and why the public/private balance between forest laws and forest certification changes as the private standards are revised and as new state policy targets are introduced (for example, with respect to biofuels and carbon capture). As certification programs mature, there will also be increasing questions about certification durability. Will certification legitimacy continue to strengthen, or will certification systems become less stable as the programs gain institutional capacity and potentially lose rule-making flexibility through industry and/or government capture? Will certification programs seek to establish a gold standard or maintain a more inclusive baseline performance bar? Organizational life cycle studies will be critical to the future design of adaptive co-regulatory systems.

As CSR standards continue to gain acceptance in resource management areas and industry sectors besides forestry, there will be more and more opportunities to conduct investigations and comparative studies of the challenges, limits, benefits, and optimal role for governments in effectively managing and leveraging private CSR initiatives. Overall, the hope is that the results and analytical framework of this book will guide and encourage future investigations into CSR co-regulation, an important new governance approach towards achieving sustainability solutions.

Appendix A
Research Interviews

Table A.1

Canada interviews

Sector	Organization	Interviewee	Date (mm/dd/yy)
Industry	Abitibi-Consolidated	Guy Tremblay	03/02/05 03/23/05
Consulting	Abusow Consulting	Kathy Abusow	06/04/04 04/11/05
Provincial government	BC Forest Practices Board	Chris Mosher	02/10/05* 03/29/05
Provincial government	BC Ministry of Forests	Jon O'Riordan*	01/14/05
		Don Wright*	02/01/05
		Johanna Den Hertog*	02/08/05
		David Morel	02/14/05
Industry	Bowater	Pierre Côté	03/02/05
ENGO	Canadian Parks and Wilderness Society (CPAWS)	Chris Henschel	03/23/05
Industry	Canfor	Peter Bentley	11/22/04
		Ken Higginbotham	01/11/05
		Lee Coonfer	01/11/05
		Paul Wooding	01/21/05
Industry	Domtar	Keith Ley	03/02/05 03/10/05
		Bernard Senécal	03/18/05
Federal government	Environment Canada	Sandy Scott	06/09/04
		Desmond Fitz-Gibbon	08/20/04
		Adam Auer	08/20/04
		Andrea Moffat	08/20/04
Industry	Forest Products Association of Canada	Andrew DeVries	06/04/04

►

◄ *Table A.1*

Sector	Organization	Interviewee	Date (mm/dd/yy)
Standards body	FSC Canada	Jim McCarthy	03/08/05
Federal government	Industry Canada	Louise Bergin John Dauvergne	04/27/04 06/04/04 06/09/04
Industry	Interfor	Rick Slaco	03/03/05
Industry	J.D. Irving	Scott MacDougall	02/17/05
Academic	Laval University	Luc Bouthèlier	03/11/05
Industry	MacMillan Bloedel	Bill Cafferata*	02/10/05
Consulting	Moresby Consulting	Patrick Armstrong	02/14/05
Federal government	Natural Resources Canada	Randall Nelson	04/27/04 04/02/04 03/02/05
Industry	New Brunswick Forest Industry Association	Yvon Poitras	02/14/05
Provincial government	New Brunswick MNR	Doug Mason	02/09/05
Industry	Office Depot	Tyler Elm	03/25/05
Provincial government	Ontario Ministry of Natural Resources	Celia Graham Betty Vankerkhof	02/04/05 03/02/05
Auditor	PricewaterhouseCoopers	Bruce Eaket	01/19/05
Provincial government	Quebec Ministry of Natural Resources, Wildlife and Parks	Germain Paré Jean Legris	04/16/05 04/16/05
Industry	Quebec Wood Export	Carl-Éric Guertin	03/14/05
Academic	Simon Fraser University	Michael Howlett	01/12/05 01/31/05
Industry	Tembec	Mike Martel	03/02/05
Industry	UPM-Kymmene	Jennifer Landry	02/18/05
Industry	Weldwood	Don Laishley* Don Wright*	11/26/04 02/01/05
Industry	West Fraser	Al Bennett	03/02/05
NGO/ Industry	MacMillan Bloedel/ Weyerhaeuser	Linda Coady*	11/29/04

* Interviews with individuals who were formerly employed by these organizations.

Table A.2

US interviews

Sector	Organization	Interviewee	Jurisdiction	Date (mm/dd/yy)
Industry	Boise Cascade	Brad Holt	Idaho	10/12/06
Industry	Bowater	Barry Graden	US South	10/30/06
Industry	Canfor Corporation	Paul Wooding	North America	10/23/06
Industry	Domtar	Jim Rodd	Wisconsin	11/02/06
State	Florida Division of Forestry	Mike Long	Florida	05/23/07
State	Indiana DNR	John Seifert	Indiana	06/27/07
Industry	International Paper	Sharon Haines	US	10/03/06
State	Massachusetts DCR	Jim Dimaio	Massachusetts	12/13/06
State	Maine Department of Conservation	Donald Mansius Tom Charles	Maine	10/20/06 10/13/06
Industry	Maine SIC	Pat Sirois	Maine	10/25/06
Certifier	Maine Master Logger Program	Sandy Brawders	Maine	10/31/06
Industry	MeadWestvaco	Joe Lawson	US	11/13/06
State	Michigan DNR	Dennis Nezich Cara Boucher	Michigan	01/16/07 01/16/07
Industry	Michigan Forest Products Council	George Berghorn	Michigan	12/05/06
State	Minnesota DNR	Andrew Arendts Tom Baumann	Minnesota	10/24/06 11/03/06
Industry	Minnesota Forest Industries Association	Tim O'Hara	Minnesota	10/23/06
State	NC Division Forest Resources	Michael Chesnutt Hans Rohr	North Carolina	10/11/06 10/11/06
State	New York Department of Environmental Conservation	Frank Dunstan David Forness	New York	07/06/07 10/24/06
Academic	North Carolina State University	Fred Cubbage	US	10/19/06
State	Oregon Department of Forestry	Marvin Brown David Morman	Oregon	10/19/06 11/03/06

►

◄ *Table A.2*

Sector	Organization	Interviewee	Jurisdiction	Date (mm/dd/yy)
NIPF	Oregon Small Woodlands Association	Mike Gaudern	Oregon	05/16/06
State	Pennsylvania DNR	Dan Devlin	Pennsylvania	10/13/06
NGO	Pinchot Institute	Will Price	US	06/11/07
		Al Sample		06/11/07
Industry	Plum Creek Timber Company	Jim Kranz	Rocky Mountain	10/20/06
		Rob Olszewski	US	10/26/06
Auditor	Pricewaterhouse Coopers	Bruce Eaket	North America	04/04/06
		Don Taylor	US	05/03/06
Auditor	SCS	Robert Hrubes	North America	10/31/06
Industry	Seven Islands	Mike Dan	Maine	11/20/06
Standards body	SFI	Bill Banzhaf	North America	10/16/06
				06/12/07
Auditor	SmartWood	Richard Donovan	US	11/09/06
County	St. Louis County	Mark Reed	Minnesota	10/18/06
Industry	Stora Enso	Gordy Mouw	Wisconsin	10/19/06
NIPF	SWOAM	Tom Doak	Maine	10/31/06
State	Tennessee Division of Forestry	David Todd	Tennessee	10/23/06
		Paul Deizman		10/26/06
NGO	The Nature Conservancy	Fran Price	US	06/12/06
Auditor	The PlumLine	Bill Rockwell	US	06/06/06
		Charles Levesque	US Northeast	10/16/06
Industry	TIME Inc.	David Refkin	Global	05/31/07
Academic	University of Minnesota	Tom Koontz	US	06/06/06
Federal	US Forest Service	Mike Higgs	US	05/16/06
		Denise Ingram		03/29/07
		Doug MacCleery		06/11/07
NIPF	Washington Farm Forestry Association	Rick Dunning	Washington	05/04/06
State	Washington DNR	Craig Partridge	Washington	10/17/06
Industry	Weyerhaeuser	Cassie Phillips	North America	03/15/06
		Kirk Titus	Minnesota	10/03/06
		Jim James	US	03/29/07
		Bob Emory	US South	10/12/06
State	Wisconsin DNR	Bob Mather	Wisconsin	10/25/06
Academic	Yale University	Connie McDermott	US	06/07/06

Table A.3

Sweden interviews

Sector	Organization	Interviewee	Date (mm/dd/yy)
Academic	FNI, Oslo	Lars Gulbrandsen	09/21/07
Standards body	FSC Sweden	Peter Roberntz (former Executive Director)	09/18/07
		Karin Fallman (Vice-Director)	09/18/07
Standards body	PEFC Sweden	Folke Stenstrom (former Executive Director)	09/11/07
		Magnus Norrby (National Executive Secretary)	09/17/07
Industry	SCA	Mårten Larsson (Manager Technical Development/TQM)	10/31/07
Academic	SLU, Uppsala	Dr. Fred Ingemarson	09/17/07
Academic	SLU, Uppsala	Dr. Matts Nylinder	09/12/07
Industry	Stora Enso	Ragnar Friberg (Senior VP Sustainability)	09/12/07
Industry	Sveaskog	Olof Johansson	10/31/07
National government	Swedish Environmental Management Council	Peter Nohrstedt (EKU Manager)	11/23/07
Small forest owners	Swedish Federation of Family Forest Owners	Jan-Åke Lunden (chief forester LRF)	09/19/07
		Dr. Tage Klingberg (former chairman, 1993-99; professor, University College of Gävle)	09/11/07
National government	Swedish Forest Agency (SFA)	Erik Sollander (Senior Advisor)	09/13/07
		Bo Wallin (former head of Environment Department)	09/14/07
Industry	Swedish Forest Industries Association	Roland Palm	09/12/07

►

◀ *Table A.3*

Sector	Organization	Interviewee	Date (mm/dd/yy)
ENGO	Taiga Rescue Network	Karin Lindahl (founding member of Taiga Rescue Network and member of FSC International board)	09/17/07
National government	SEPA	Sune Sohlberg	11/01/07
Academic	UBC, Vancouver	Dr. Gunilla Öberg	08/29/07
Academic	Umeå University	Dr. Katarina Eckerberg	06/07/06
ENGO	WWF, Swedish Chapter	Lena Dahl (coordinator, TetraPak; former coordinator of WWF-Sweden Forest and Trade Network)	09/13/07

Appendix B
The Leading Global Forest Certification Programs

Forest Stewardship Council (FSC)

The FSC is an international nonprofit NGO. Founded in 1993, it is run by a board of environmental, business, and social interests. The FSC is a membership organization with nearly 600 members from more than seventy countries. Its mission is to promote environmentally appropriate, socially beneficial, and economically viable management of the world's forests according to ten FSC principles and fifty-six criteria. A company certifies to the relevant regional standard. The FSC standard includes a chain-of-custody certification and label.

See http://www.fsc.org.

Programme for the Endorsement of Forest Certification (PEFC)

The PEFC is an independent international nonprofit NGO established in 1999 to provide an umbrella framework for the development and mutual recognition of national or subnational forest certification programs (under a common eco-label) that meet internationally recognized requirements for sustainable forest management. Originally founded by private landowners in Europe to accredit their national and regional forestry certification programs, the PEFC membership now includes approximately thirty endorsed independent national forest certification programs worldwide.

See http://www.pefc.org.

Canadian Standards Association Sustainable Forest Management Standard (CAN/CSA-Z809)

The CAN/CSA-Z809 standard is a national SFM standard developed through a multi-stakeholder process under the auspices of the Canadian Standards Association, an independent nonprofit organization accredited by the Standards Council of Canada (SCC). The standard was first published in 1996. It sets requirements to be met under six broad nationally and internationally recognized SFM criteria. SFM performance requirements are defined through an ongoing local public participation committee and complemented by management system requirements consistent with the ISO 14001 standard. Companies seeking to certify a defined forest area (DFA) through the CSA must undergo an independent third-party audit of their management system and a field inspection to confirm the attainment of performance objectives. In 2001, the CSA launched a chain-of-custody and labelling option. The CAN/CSA-Z809 standard was endorsed by the PEFC in March 2005.

See http://www.csasfmforests.ca.

Sustainable Forestry Initiative (SFI)

The American Forest and Paper Association (AF&PA) developed the SFI standard in 1994. In 2000, an independent Sustainable Forestry Board (SFB) was established to oversee the SFI standard's ongoing development. The SFB included representatives from industry and from the environmental, conservation, academic, and public sectors. The SFI standard includes a set of SFM principles, objectives, performance measures, and core indicators. It also offers a certified procurement system audit and an on-product label option. The 2005-09 SFI revised standard requires SFI auditors to be accredited by the American National Standards Institute (ANSI) or the Standards Council of Canada (SCC). In January 2007, the SFI separated from the AF&PA and became a fully independent non-profit organization (SFI Inc.) with a fifteen-member multi-stakeholder board of directors. The PEFC endorsed the SFI program in December 2005.

See http://www.sfiprogram.org.

The American Tree Farm System (ATFS)

The ATFS program applies to non-industrial forest owners in the US. The ATFS has existed since 1941 but certification standards were approved only in 1998. Those seeking certification must have a written management plan based on the ATFS SFM standards and guidelines. Accredited auditors conduct the independent certification inspections. The ATFS was endorsed under the PEFC program in August 2008.

See http://www.treefarmsystem.org.

Appendix C
Summary of US State Forest Agency Interviews

The following summarizes the state forest agency interviewee responses to three questions:

- What were the key drivers that led your state to certify its state forests when you did?
- What were the implementation challenges?
- What have been the outcomes from certifying your state forests?

It should be noted that interviewee responses were to a set of open questions rather than to a formal pre-structured questionnaire.

Table C.1

	PA	NC	TN	ME	NY	MA	MN	WI	MI	WA	IN
Drivers											
Pinchot funding	✓	✓	✓	✓	✓		✓	✓			
Buyer pressure				✓			✓	✓	✓		
ENGO advocacy			✓	✓					✓	✓	✓
State economy				✓			✓	✓	✓		✓
Interstate competition							✓	✓	✓	✓	✓
State leadership	✓	✓		✓		✓	✓	✓			
Market opportunity	✓		✓	✓	✓			✓			✓

►

◄ *Table C.1*

	PA	NC	TN	ME	NY	MA	MN	WI	MI	WA	IN
Challenges											
Workload and documentation		✓	✓	✓	✓	✓	✓	✓	✓		
Budget justification		✓	✓	✓	✓	✓	✓		✓	✓	
Coordination and policy alignment	✓					✓	✓	✓	✓	✓	
Public sector flexibility	✓	✓	✓		✓			✓	✓	✓	
Gaining staff cooperation				✓	✓	✓	✓	✓	✓	✓	✓
Addressing SFM audit findings		✓			✓	✓	✓		✓		
Outcome benefits											
Transparency and accountability		✓		✓		✓		✓		✓	✓
State forest administration	✓	✓	✓	✓	✓	✓	✓	✓	✓		✓
State forest management		✓		✓		✓		✓	✓		✓
Market gains				✓			✓			✓	✓
State leadership	✓			✓	✓	✓	✓	✓	✓		✓

Appendix D
US State Forest Certification Audit Outcomes

As reported in the respective state forest certification audit summary reports, forest governance improvements resulting from state forest certification included those in the following table. The state forest certification summary audit reports can be obtained from the respective state forest agency websites, as well as from the Sustainable Forestry Initiative, Forest Stewardship Council, and certification auditor web pages. For example, see Sustainable Forestry Initiative, http://www.sfiprogram.org/forest_certification_audits_reports.cfm; Scientific Certification Systems, http://www.scscertified.com/forestry/forest_certclients.html; and Rainforest Alliance, http://www.rainforest-alliance.org/forestry/public_documents.cfm.

Table D.1

Audit outcomes

State	Audit outcomes
Maine	• Land management planning and policies have been improved, including improved land classification; sustainable harvest levels; increased focus on wildlife, biodiversity, and landscape-level issues; riparian management standards; and water quality and habitat management. • Harvest practices with respect to wildlife habitat and aesthetic considerations and clearcut implementation have been improved. The overall best management practice (BMP) effectiveness is at 82%. • Identification of operations out of compliance with BMPs/regulations has been improved, and there are better processes to help correct behaviour. • The definition of old growth has been revised. • The policy on legacy and reserve trees has been revised.

►

◄ *Table D.1*

State	Audit outcomes
Maryland	• Modelling software has been acquired and collaborative work initiated between Department of Natural Resources (DNR) and Heritage staff. • Staff job descriptions have been updated to include performance of Sustainable Forestry Initiative (SFI) and Forest Stewardship Council (FSC) requirements. • A sustainable forestry information sheet has been developed for loggers and timber sale contracts. • A task group has been established to define and map high conservation value forests (HCVFs). • A *Sustainable Forest Management Plan* and *Annual Chesapeake Forest* summary report have been prepared and posted on the website. A *Summary of the Chesapeake Forest Monitoring Plan* report has also been posted on the DNR website. • Citizen Advisory committees have been reorganized (combined) to improve efficiency.
Michigan	• A Timber Pre-sale Checklist has been implemented. • Additional funding has been made available to address identified issues such as BMP follow-up and Outdoor Recreation Vehicle (ORV) trail improvement (e.g., ORV Task Force created), and increased resources have been made available for management plan updates. • There is increased DNR involvement on the State SFI Implementation Committee. • DNR staff working with the Office of the State Archeologist to develop staff training on site identification and reporting. • Stand retention guidelines have been developed. • Timelines have been established and met for completion of state forest plans. • A tracking and reporting system has been implemented to identify water quality and soil erosion issues and make the case for funding to fix the problems.
Minnesota	• DNR discontinued the use of simazine as of December 2005. • Thresholds have been established for residual stand damage and rutting according to DNR site-level guidelines. • Statutory requirements have been enacted regarding required logger BMP and safety training. • An elaborate monitoring program has been developed and an internal audit team established for tracking and following up on management plans. • Certification is now in the DNR vernacular; e.g., loggers are now talking about corrective action requests that they have to address.

►

◀ *Table D.1*

State	Audit outcomes
North Carolina	• There is a policy to limit clearcut harvests in plantations to less than 40 acres without green tree retention. • A geographic information system (GIS) has been established. • There are improved methods for handling records of forest monitoring and management activities and reporting on results. • A collaborative biological survey has been undertaken with NC Natural Heritage Program, and a cultural survey initiated with NC Department of Cultural Affairs. • A formal policy has been established for stakeholder contact and communication, and a procedure for public input into management planning has been implemented. • Identification and mapping of high conservation values in Bladen Lakes State Forest has been done. • Training has been conducted for all field personnel on management plan objectives, processes, and procedures.
Pennsylvania	• Contracts now include a safety clause, and contracts are being used consistently across the Bureau of Forests (BOF) Districts. • A Deer Management Plan for Pennsylvania's state forestlands and Department of Conservation and Natural Resources (DCNR) Action Plan (2004-05) for deer management have been developed. • The DCNR and the Pennsylvania Game Commission are working cooperatively on deer management research and implementation of the Action Plan. • A landscape examination and planning method has been developed. • A logger certification requirement has been added to timber sale contracts. • A system for training, guiding, and supervising logging contractors is in place and documented. • A training database is being developed for foresters and Forest Division staff. • Road inventories are conducted and indicators established for assessing environmental impacts.
Tennessee	• Safety and forest management training plans have been developed and implemented. • New positions have been created and salaries improved (e.g., the positions of State Forest Supervisors and "Outreach/Information and Education Unit Leader" have been established).

▶

◄ *Table D.1*

State	Audit outcomes
	• New policies and procedures have been developed and communicated through the convening of a forum on state forest practices. • A "Framework for State Forest Resource Monitoring" has been developed and implemented. • A state forest inventory plan has been developed and is being implemented. • A planning process to identify the need for and extent of representative ecosystems and high conservation value forests has been developed. • Specific and achievable objectives have been included in all revised management plans. • A State Forest Monitoring Plan has been developed and implemented through inter-agency coordination (e.g., with the Department of Environmental Conservation). • There are further coordination opportunities with groups such as the Conservation Heritage Foundation, Nature Conservancy, Conservation Commission, and Tennessee Wildlife Resources Association.

Notes

Introduction

1 See Anderson 1989 for a concise historical overview of CSR debates regarding the role and responsibilities of commercial entities dating back to the pre-medieval period, through the mercantile period and early industrial era, to the present day.

2 The term "sustainable forest management" (SFM) is employed throughout this book. As a policy goal, it refers to the balancing of economic, environmental, and social forest values so as to ensure a healthy and productive forest landscape that can meet the needs of present and future generations.

3 In other words, the authority "pie" is only so big, so if private authority increases, then public authority must decrease – that is, the state must be in retreat.

4 Co-regulation in a general sense refers to shared decision making. In this book, I define co-regulation as cooperative policy making between state and non-state private actors (including corporations and environmental advocacy organizations).

5 For a concise summary of forest certification in Finland, see Cashore et al. 2007.

6 High *public* capacity refers to regions with well-established legal frameworks and forest institutions. High *private* capacity concerns global forest producing regions with multinational forest company ownership and/or management. It should be noted that there is a spectrum of forest governance capacity across developing regions. For example, some lesser-developed states have strong regulations but insufficient enforcement, whereas others have a fundamental lack of adequate forest rules and regulatory structures.

7 In this book, developed countries are distinguished from developing countries by their membership in the Organisation for Economic Co-operation and Development (OECD).

8 Although it is a significant Canadian forest-producing region, I exclude the province of Alberta since forestry constitutes only a small percentage of the overall provincial economy relative to oil and gas. The Alberta government has played a hands-off role in certification, addressing forestry within integrated land management decisions.

9 Although elected officials play an obvious role in supporting forest policy initiatives, certification co-regulation has been largely non-partisan, with policies and programs developed and delivered at the level of the bureaucracy and carried forward by both left- and right-of-centre governments across electoral cycles.

10 In addition, although SFM encompasses both social as well as environmental considerations, I concentrate on the environmental aspects of sustainability rather than social considerations such as equity, security, employment, and community health and safety. Community engagement is addressed within the context of encouraging improved forestry practices to achieve and maintain healthy and productive forests.

Chapter 2: Co-Regulating Corporate Social Responsibility

1 The terms "corporate responsibility" and "corporate sustainability" are often used interchangeably with the concepts of CSR and corporate citizenship to describe the social and

environmental responsibilities of the firm beyond legal compliance and maximization of shareholder profit. This book uses the term "corporate social responsibility" (CSR) to encompass all of these terms. The foundational literature on corporate citizenship and CSR includes Bowie 1991; Carroll 1991, 1999; Elkington 1998; and Zadek 2001.

2 For the foundational literature on sustainable development, see Dale and Robinson 1996; Daly 1990; Meadowcroft 2000; World Commission on Environment and Development 1987; and World Business Council for Sustainable Development 2001. For an understanding of ecological modernization theory (i.e., sustained growth through "green" technological innovation), see Christoff 1996; Hajer 1995; Mol and Sonnenfeld 2000; and Spaargaren and Mol 1992.

3 For literature on public administration efficiency reforms, see Hood 1991; Osborne and Gaebler 1993; Peters 1994; and Sabatier 1986.

4 For literature on transnational advocacy group governance, see Keck and Sikkink 1998; and Keohane 2003.

5 See Gunningham 2007, 481-85, and Gunningham and Sinclair 2002a, 135-36, for an explanation of the importance of "social licence," i.e., meeting societal expectations to maintain corporate privileges (beyond legal and economic licence to operate).

6 The World Bank has estimated that approximately 1,000 codes of conduct have been developed by multinational firms across a range of sectors, including apparel, footwear, agribusiness, tourism, and the oil and gas and mining resource sectors. See World Bank 2003. For an inventory description of industry codes of conduct, see OECD 1999.

7 In addition to environmental standards, a range of social codes, labelling schemes, and certification systems concerning child labour and working conditions emerged during this period, including the Sullivan Principles, the Social Accountability 8000 CSR standard, and the RugMark and Fair Trade coffee social labelling programs. For a description of the range of global CSR codes, labels, and standards, see Conroy 2007; Leipziger 2003; and McKague and Cragg 2007.

8 For literature on the shift from government to governance, see Kooiman 1993; Mayntz 2003; Peters and Pierre 1998; Rhodes 1996; Rosenau and Czempiel 1992. For more recent literature on the transformed state role, see Bartle and Vass 2005; Hennebel et al. 2007; Heritier 2001; Jordan, Wurzel, and Zito 2005; Knill and Lehmkuhl 2002; and Schulz and Held 2004.

9 "Rule making" refers to the formulation of regulations. "Implementation" concerns the on-the-ground delivery of the rules. "Enforcement" refers to the mechanism for ensuring transparency and accountability of rule implementation. This typology is based on the "stages heuristic" developed by policy scholars Jones, Anderson, Brewer, and deLeon, as reviewed by Sabatier 1999, 6-8.

10 "Soft law" is declaratory but non-binding law. Examples include communication, knowledge transfer, and voluntary approaches such as industry self-regulation, voluntary codes, voluntary challenges, charters, covenants, and negotiated agreements.

11 For literature on scaled, progressive regulatory enforcement, see Ayres and Braithwaite 1992; Gunningham and Grabosky 1998; Gunningham and Sinclair 1999; Phidd and Doern 1983; Webb 2005.

12 "Meta-governance" refers to the governance of governance – for example, state governance of self-regulation. See Peters 2006; Sørensen and Torfing 2007; and Webb 1999, 2005. See also Parker 2007 for a discussion of meta-regulation (i.e., employing the law to encourage beyond-compliance CSR behaviour). For network management literature, see Jessop 2002; Kickert, Klijn, and Koppenjan 1997; Kooiman 1993, 2003; and Koppenjan and Klijn 2004.

13 For CSR, see Auld, Bernstein, and Cashore 2008; Moon 2002b; Vogel 2005. For NSMD, see Cashore 2002; Cashore, Auld, and Newsom 2004. For non-state global governance, see Bernstein and Cashore 2004. For private regulation, see Bernstein and Cashore 2007; Meidinger 1999. For private hard law, see Cashore et al. 2007. For civil regulation, see Meidinger 2003b; Vogel 2006. For codes of conduct, see Jenkins 2001.

14 For studies on co-regulation (the policy and practice of co-regulation), see Bartle and Vass 2005; European Economic and Social Committee 2005; Eijlander 2005; Hennebel et al. 2007; Palzer and Scheuer 2004; Senden 2005.

15 This differs slightly from the EU definition, which regards co-regulation as a mechanism for implementing legislation through delegated self-regulation. See Palzer and Scheuer 2004. Specifically, the EU defines co-regulation as "the mechanism whereby a Community Legislative Act entrusts the attainment of the objectives defined by the legislative authority to parties which are recognized in the field (such as economic operators, the social partners, non-governmental organizations, or associations)." See European Commission 2001. The term "co-regulation" is less commonly applied to describe the joint private governance arrangements between regulated organizations (such as corporations) and nongovernmental actors (such as civil society organizations). See Pattberg 2007; Utting 2005. In some instances, the term "co-regulation" is also employed to describe joint governance arrangements between government authorities (collaborative governance). See Ansell and Gash 2008.

16 These scholars argue that the state is in retreat as state sovereignty and capacity to govern at both the domestic and international levels has receded relative to accelerating financial and technical resources of private actors, global networks, and increasing private self-regulatory governance authority. See Ohmae 1995; Strange 1996.

17 In the global governance literature, "hollowing of the state" refers to "the state losing power and authority upward to supranational institutions; sideways to the private sphere; and downward to the increasing demands for localism and devolved government." See Lister and Marsh 2006, 258. In the public policy literature, the "hollowing of government" refers to new public management reforms that transfer various functions and activities traditionally undertaken by governments to private actors. See Brinton and Provan 2000; Brinton, Provan, and Else 1993; Howlett 2000; Peters 1994.

18 See Mayntz 2003; Peters and Pierre 1998; Rhodes 1996; Rosenau and Czempiel 1992.

19 See Gunningham and Sinclair 1999; Karkkainen 2004; Knill and Lehmkuhl 2002; Sørensen 2004.

20 The term "regulatory capitalism" refers to the expansion in the scope (state and non-state), arenas (international and domestic), instruments (hybrid), and depth of regulation (more governance of more kinds). See Levi-Faur 2005; Braithwaite 2008.

21 For studies on the strengths and weaknesses of voluntary policy approaches, see Coglianese and Nash 2001; Gibson 1999; Gunningham and Rees 1997; Harrison 2002b; Kagan, Gunningham, and Thornton 2003; May 2005; Morgenstern and Pizer 2007; OECD 2003; Potoski and Prakash 2002; Webb and Morrison 2005.

22 For literature on the effectiveness of voluntary instruments, see Antweiler and Harrison 2007; Harrison 2002a; Lyon and Maxwell 2002; Morgenstern and Pizer 2007; OECD 2003.

23 The authors define governance capacity as the formal and factual capability of public or private actors to define the content of public goods and to shape the social, economic, and political processes by which these goods are provided.

24 Besides Ayres and Braithwaite 1992; Gunningham and Rees 1997; Haufler 2003; and Knill and Lehmkuhl 2002, for literature on the new tools of governance, see Eliadis, Hill, and Howlett 2005; Howlett 2000, 2005; Jordan, Wurzel, and Zito 2003, 2005; and Salamon 2002.

25 For example, they highlight the fact that information can be a complement to all instrument categories. They argue that unilateral voluntary initiatives have serious flaws as a stand-alone policy instrument (e.g., free-riding) that can be compensated by underpinning regulation. Finally, in terms of sequencing, more coercive regulatory approaches should be adopted only if softer policy instruments have failed to change behaviour. See Gunningham and Grabosky 1998, 422-44.

26 Effectiveness concerns the realization of outcome goals. Efficiency refers to the input/output ratio of policy implementation resources. Legality concerns the correspondence with formal rules and principles of due process. And "democratic values" refers to the correspondence of instrument design and implementation with accepted norms of citizen-government relationships in a democratic political order.

27 For literature regarding the legitimacy of private environmental governance systems, see Bernstein 2005; Bernstein and Cashore 2007; Cashore 2002; Cashore, Auld, and Newsom 2004; Dingwerth 2005; Wolf 2006.

28 For literature regarding the accountability of global private environmental governance standards, see Boström 2005; Gulbrandsen 2008a; Held and Koenig-Archibugi 2005; Keohane 2003.

29 For literature regarding evaluating the effectiveness of transnational private environmental governance systems (private environmental regimes), see Miles et al. 2002; Mitchell 2002; Newell 2001; Sprinz and Helm 1999; Wettestad 2001; and Young 1999, 2001.

30 Output effectiveness concerns the development and uptake of a relevant policy tool. Outcome concerns behavioural change, and impact criteria are the on-the-ground consequences of the policy. Output and outcome effectiveness are preconditions of impact effectiveness. See Easton 1965.

31 For explanations of the statist perspective, see Gereffi, Garcia-Johnson, and Sasser 2001; Kahn and Minnich 2005; Koenig-Archibugi 2005, 132; Lipschutz and Rowe 2005.

32 For studies on CSR and development, see Calder and Culverwell 2005; Fox 2004; Fox, Ward, and Howard 2002; Nelson 2006; Newell 2002; Ruggie 2004; Swift and Zadek 2002; Utting 2003; Ward 2004.

33 For literature and reports on government's role in CSR in the case of developed nations, see Albareda, Lozano, and Ysa 2007; Bartle and Vass 2005; Bichta 2003; Bleishwitz 2003; Carey and Guttenstein 2008; European Commission 2006; GAO 2005; Hennebel et al. 2007; Moon 2002a; Natural Resources Canada 2004b; Zappala 2003.

34 A compendium of national policies supporting CSR within EU countries is found at http://ec.europa.eu/enterprise/policies/sustainable-business/. The US Government Accountability Office (GAO) prepared a summary of US federal policies and programs supporting CSR. See GAO 2005.

35 For example, the Canadian federal government (Industry Canada) explains that they promote CSR principles and practices to Canadian businesses because it helps firms be more competitive by supporting operational efficiency gains, improved risk management, favourable relations with the investment community, improved access to capital, enhanced employee relations, stronger relationships with communities and an enhanced licence to operate, and improved reputation and branding. See http://strategis.ic.gc.ca/epic/site/csr-rse.nsf/en/Home.

36 For example, the World Bank's Business Competitiveness and Development Program, the Kennedy School of Government's CSR Initiative, the Keenan Institute, the Prince of Wales International Business Leaders Forum, the European Academy of Business in Society (EABIS), and the Responsible Competitiveness Consortium have all initiated research efforts to examine the role of public sector capacity in enabling CSR.

37 See http://www.ifc.org/ifcext/economics.nsf/Content/CSR-diagnostic.

38 "Greenwash," as defined in the *Oxford English Dictionary* refers to "disinformation disseminated by an organization so as to present an environmentally responsible image."

39 NGO and civil society advocates of government intervention in CSR argue that there is a need to mandate CSR in order to speed up implementation and increase CSR scale. See Christian Aid 2004; SustainAbility 2004. In addition, there is support from the business community, which sees CSR legislation as a way to level the playing field, provide clear direction among the array of voluntary codes, and define clear CSR boundaries with regard to public versus private responsibilities. See World Economic Forum 2008.

Chapter 3: Government's Role in Forest Certification

1 Deforestation rates have slowed in the temperate forests of Europe, China, and North America due to reforestation and afforestation efforts, as well as a logging ban in China. The Chinese government banned logging in natural forests in 1999 after extensive and devastating floods on the Yangtze River in 1998, partially attributing the floods to logging in the headwaters of the river.

2 In the period 2000-05, 55 percent of global deforestation occurred in only 6 percent of the world's forests. Of the world's total forest loss, 48 percent is occurring in Brazil and 13 percent in Indonesia. African forests are also being deforested but account for only 5.4 percent of the world's total forest loss. See Hansen et al. 2008.

3 Terms such as "ancient," "frontier," "endangered," "old growth," and "high conservation value" forest are often used interchangeably to refer to primary forests that are undisturbed by human activity. The terms have different, politically derived meanings to different organizations, however. Fundamentally, they were coined for communications and advocacy purposes rather than necessarily derived from forest science research.

4 In British Columbia, there were ongoing valley-by-valley contests (e.g., Stein Valley, Clayoquot Sound, Great Bear Rainforest, etc.) between forest companies and ENGOs over the clearcut logging of old-growth forests. See Coady 2002. In the US Pacific Northwest, heightened controversy over logging in spotted owl habitat (in old-growth forests) resulted in the virtual shutdown of harvesting on US national forestlands. In Sweden, advocacy campaigns were launched against forest companies such as Störa to halt the logging of the country's northern "ancient forests." See Gamlin 1988.

5 There have been many state-led international cooperative initiatives that have also been unsuccessful in curbing tropical forest degradation (e.g., the Tropical Forestry Action Plan, conservation concessions, and debt-for-nature swaps). The critical focus here, however, is on the failed attempts to establish *binding* international forest law.

6 In 1990, prior to UNCED, the WWF had also attempted and failed with a proposal to the International Tropical Timber Organization (ITTO) to establish an independent scheme to assess and certify tropical forest sustainability. The proposal was a follow-up to ITTO's 1985 pledge to trade only in forest products from sustainably managed forests by 2000. See Humphreys 1996.

7 The FSC defines consensus as the absence of sustained opposition but does not require unanimity. In the case of a vote, "decisions shall require both the affirmative vote of a simple majority of members within each sub chamber, and 66.6 percent of the total voting power registered by Associates in good standing (calculated as provided for in these By-laws) with exception of Board elections" (FSC-AC 2006).

8 Plantation management was added as a principle in 1996.

9 Several governments, including Austria, the Netherlands, the United Kingdom, Mexico, and Switzerland have provided funding for the FSC. The Austrian government provided most of the funding over the first two years. They redirected approximately US$1.2 million from a rescinded law/program that had been intended to restrict the import of tropical timber but had been challenged as a barrier to trade under the General Agreement on Tariffs and Trade (GATT).

10 In this context, "national certification programs" refers to competitor programs to the FSC and not to the FSC's national and subnational regional standards.

11 Some studies have steered around these challenges by examining the certification audit corrective action requests to determine the extent of forest practice improvements. See Newsom, Bahn, and Cashore 2006; Rametsteiner and Simula 2003; Tikina 2008; and WWF 2005. Others have surveyed forest managers regarding their perceptions of the degree of change in forest practices resulting from certification. See Tikina et al. 2009. In addition, McDermott and colleagues (2008) assessed forest certification effectiveness by comparing private forest rules with prescriptive legislated requirements within and between political jurisdictions.

12 Studies examining certification effectiveness in terms of successful uptake include Ebeling and Yasue 2009; Espach 2006; and Gulbrandsen 2005b.

13 For example, see Rickenbach and Overdevest 2006 for a study of the expectations of US FSC-certified organizations that confirms the role of market signalling and learning as key certification drivers along with financial market incentives.

14 For example, ENGOs are continuing their market campaigns. In the US, ForestEthics is targeting catalog producers such as Victoria's Secret and Sears/Lands' End to stop sourcing fibre from endangered forests (e.g., the Canadian boreal forest) and establish procurement policies requiring FSC certification.

15 For example, the California Climate Action Registry (CCAR) allows only carbon sequestered from certified SFM forests to be registered in its carbon listing.

16 For the sociological theory regarding legitimacy, see Suchman 1995.

17 For comparisons of the certification standards and evaluation frameworks, see Abusow 2001; CEPI 2004; Commonwealth of Australia 2000; CPET 2006; FPAC 2006; Heaton 2001; Meridian Institute 2001; and WWF/World Bank 2006. For a good summary review of the various efforts, see Nussbaum and Simula 2004.
18 For critiques of the PEFC, see FERN 2004; Roberts 2007; WWF 2001.
19 In 2001, Greenpeace and WWF rejected a proposal by the International Forest Industry Roundtable to create an international framework for mutual recognition of PEFC and FSC programs.
20 For example, in 2004 the British government established the Central Point of Expertise on Timber Procurement (CPET) to assess the five internationally recognized certification schemes and provide guidelines for central government departments on legal and sustainable timber procurement. In the first assessment, all five forest certification schemes were found to meet UK government requirements for legality, but only two (FSC and CSA) met the requirements for sustainability. In April 2005, SFI and PEFC were reassessed and accepted as proof of legal and sustainable timber. See CPET 2006. Also see http://www.cpet.org.uk.
21 "Mutual recognition" criteria judge whether an individual certification scheme is credible, whereas "legitimacy threshold" criteria differentiate between the credibility of the various attributes of the various standards.
22 An additional aspect of the credibility debate is that disagreements have arisen within the FSC itself. Some traditional FSC supporters are protesting that the FSC has lost its credibility by awarding certification to unsustainable forestry operations. These internal contests are also providing critical feedback and encouraging ongoing adjustments and revisions to the FSC program. See http://www.fsc-watch.org and http://www.rainforestportal.org, as well as Counsell and Loraas 2002; Elad 2001; and Roberts 2007.
23 As shown in Table 3.1, various groups have initiated the leading certification systems. ENGOs led the creation of the FSC, industry associations initiated the SFI and CAN/CSA-Z809 programs, and small forestland owners established the ATFS and PEFC programs. All programs now have multi-stakeholder rule-making mechanisms and independent audit processes.
24 These distinctions have been noted by Auld, Bernstein, and Cashore 2008, 18; Bernstein and Cashore 2007, 349; and Cashore, Bernstein, and McDermott 2007.
25 For *civil regulation,* see Bendell 2000 and Meidinger 2003a. For *transnational business regulation,* see Pattberg 2006. For *supra-governmental regulation,* see Meidinger 2008. For *private hard law regulation,* see Cashore et al. 2007. For *NSMD,* see Cashore 2002 and Cashore, Auld, and Newsom 2004.
26 NSMD systems also include certification efforts in other sectors, including Fair Trade Labelling Organizations International (1997), the Marine Stewardship Council (1999), the Sustainable Tourism Stewardship Council (under development), and the Initiative for Responsible Mining Assurance (under development) (Cashore et al. 2007, 6-8).
27 Although forest certification organizations incorporate due process in terms of multi-stakeholder decision making, certification bodies are ultimately unelected and self-selected and therefore lack democratic accountability to the general public. The durability of certification as an effective governance mechanism is uncertain, as ongoing compliance can be encouraged only through the threat of certificate suspension rather than by means of a formal state-based coercive mechanism.
28 For example, in April 2007 the UN Forum on Forests (UNFF) reached another non-binding agreement on forests (the Non-legally Binding Instrument on All Types of Forests). The agreement was presented to the UN General Assembly for adoption in October 2007. The agreement focused on enhancing cooperation between countries on achieving global forest goals by 2015 and encouraging the implementation of national forest programs. Other active international forestry governance forums include the ITTO, the Ministerial Conference on the Protection of Forests in Europe, the World Trade Organization (WTO) (regarding forest trade aspects), and the World Bank (regarding forestry development project financing issues).
29 Studies examining government's role in certification include Rametsteiner 2002 and Segura 2004. For research on forest certification interaction with public policy, see Boström 2003;

Cashore and Lawson 2003; Elliott 2000; Gunningham and Sinclair 2002b; Hysing and Olsson 2005; McDermott et al. 2008; Meidinger 2003a; and Rametsteiner and Simula 2003.

30 As long as certification remains voluntary, there is no contravention of WTO rules or EU procurement directives. As governments establish public procurement policies, however, questions are arising as to whether certification becomes a technical barrier to trade. As well, as long as certification remains an NGO standard, it is not considered under WTO rules, but as governments become directly engaged in certification programs and adopt certification in procurement and state forest policies, certification may be considered a technical standard under the WTO.

31 There are forest planning, documentation, monitoring, auditing, and consultation requirements that add variable expenses as well as fixed costs per hectare of certified forestland.

32 For example, as explained by the Ghana Forestry Commission at the UNECE meeting in October 2005, "in Ghana, the forest industry is fragmented and has serious liquidity problems. In the short to medium term, forest areas may not qualify for certification because of lack of management plans, even for those forests that are legally operated. For the few that do meet certification standards, the small-scale owners are unable to afford to join a scheme." See Koleva 2006.

33 Through the EU's Forest Law Enforcement, Governance and Trade (FLEGT) process to address illegal logging, governments are now supporting a phased procurement approach towards developing regions, beginning with a requirement for "legally" supplied timber and then eventually achieving "sustainable," certified wood supply (http://www.illegal-logging.info). Industrialized countries are also encouraged to provide capacity-building resources and assistance to help develop certification in these regions.

Chapter 4: Canada

1 There was a three-year delay from 1996 to the first CSA certification in 1999, largely due to the rigorous environmental management system (EMS) documentation and local public participation elements in the CSA standard.

2 The Canadian Standards Association is a not-for-profit membership-based association that functions as a neutral third party, providing a structure and a forum for developing technical product and process standards.

3 In 2001, the Canadian Pulp and Paper Association (CPPA) changed its name to the Forest Products Association of Canada (FPAC) to broaden its focus.

4 Canada was the first country to develop a National Forest Strategy committing itself to achievement of SFM on a national level. For a description of the CCFM's SFM criteria and indicators, see CCFM 1992, 1995.

5 Section 4.12 of the National Forest Strategy states: "Industry and governments will work cooperatively to pursue joint technical discussions aimed at internationalizing product standards, codes and certification procedures." Section 4.13 states: "By 1995, industry and governments will develop and put into operation a means of identifying and promoting Canadian forest products that reflect our commitment to sustainable forests and environmentally sound technologies." See CCFM 1992.

6 These percentages are as of January 2008. In September 2008, AbitibiBowater announced its commitment to FSC-certify its 3.2 million hectares of forestland in Ontario, Quebec, and Nova Scotia.

7 Also referred to as the Ministry of Forests (MoF) and the BC Forest Service.

8 The PAS was established in 1992, committing the province to achieving protection of 12 percent of the provincial land base by the year 2000. A three-year timber supply review was initiated in 1992 to review and reset the province's allowable annual cut. The CORE was established in 1992 to develop regional land-use strategies. The LRMP consensus-building process was established in 1993 to help guide resource management objectives on Crown land at the local level.

9 Iisaak Forest Resources was established in 1998 as a joint venture between MacMillan Bloedel and local Nuu-chah-nulth First Nations: Ahousaht, Hesquiaht, Tio-o-qui-aht, Toquaht, and Ucluelet. These First Nations became the sole owners of Iisaak and Tree Farm Licence 57 in 2005.

10 Haliburton Forest in Ontario was the first formal FSC certification in Canada, in 1998. J.D. Irving's Black Brook operation in New Brunswick was certified to the FSC International principles in October 1998, but this was not formally announced. MacMillan Bloedel's North Island operation and Weldwood of Canada's 100 Mile House division dual-certified to the CAN/CSA-Z809 and ISO 14001 standards in May 1999 and December 1999, respectively. Canfor certified initially to the ISO 14001 standard in 1999, and followed this in July 2000 with CAN/CSA-Z809 certifications of its northern Vancouver Island (TFL 37) and central BC (TFL 48) tree farm licences.

11 Forest licensees in BC are allocated an allowable annual cut that is distributed across a timber supply area (TSA), resulting in a "Swiss cheese" pattern of isolated pockets of timber rather than one contiguous defined forest area.

12 In January 1998, the chairman of B&Q home improvement company announced that it would source only FSC-certified wood products by the end of 1999, i.e., they would be phasing out the sourcing of hemlock stair parts from BC, where there is reluctance to go with FSC certification. In 1998, in response to pressure from the BBC magazine, German publishers suspended their contract with Western Forest Products. In August 1999, MacMillan Bloedel's major client Home Depot announced its preference for purchasing FSC-certified wood products. In response, both MacMillan Bloedel and Western Forest Products signified their intention to consider FSC certification; neither company ended up certifying its operations with the FSC, however.

13 For a listing and description of the ecosystem-based management (EBM) goals and principles, see http://www.llbc.leg.bc.ca/public/pubdocs/bcdocs/369635/principlesgoalsebm.pdf. The JSP (now referred to as the Rainforest Solutions Project) led to the eventual signing, in February 2006, of the historic Great Bear Rainforest agreement between the province and Coastal First Nations to protect one-third of these coastal forests and to implement EBM throughout the region. See http://savethegreatbear.org/solutions/. An FSC audit of 2.2 million hectares of the mid-coast (within the Great Bear Rainforest) was conducted in November 2008.

14 Interview with Don Wright, 1 February 2005. Wright was Assistant Deputy Minister in the Ministry of Forests from 1993 to 1995, vice president at Weldwood from 1997 to 2001, and Deputy Minister in the Ministry of Forests and Range from 2001 to 2003.

15 Ibid.

16 In 2001, the BC government announced its "New Era of Sustainable Forestry" commitments, including the objective of cutting the forestry regulatory burden by one-third by 2004, without compromising environmental standards. Given provincial budget constraints, the Ministry of Forests was directed to streamline the Forest Practices Code and pursue more efficient and effective alternative arrangements for policy delivery with the forest industry and other stakeholders. In response, during 2002/03, Ministry of Forests staffing levels were cut back 14.5 percent and the ministry's budget was reduced by 15.5 percent. In 2004, the government finalized new legislation, the Forest and Range Practices Act (FRPA), to replace the Forest Practices Act (1994). Rather than enforcing highly prescriptive operational-level forest regulations, the new act focused on enabling outcome. Under the FRPA, the government set the forest conservation objectives and forest licensees were to determine how best to meet those objectives. The act emphasized much greater professional and company accountability and raised new questions about the role of certification as a policy tool to contribute to the province's forest conservation objectives.

17 Interview with Paul Wooding, Manager of Certification and Market Support, Canfor, 20 January 2005. The preamble to the guidance document for the CSA standard asserts that CAN/CSA-Z809 would "ensure that the CCFM criteria for SFM are being met."

18 Two BC government officials participated in the FSC-BC regional process, one from the Ministry of Forests and another from the Ministry of Environment, Lands and Parks.

19 For example, in 2003 the province created the Forestry Innovation Investment Ltd. agency, and under it the BC Market Outreach Network to promote BC forest products overseas.

20 Interviews with BC forest companies (see Appendix A).

21 The three pilot projects included the Morice, Fort St. John, and Kamloops timber supply areas. The government also tested the FSC system in its small business program in the

Kootenay Lake district, and also piloted an ISO 14001-based provincial environmental management system (EMS) at the Prince George, Salmon Arm, and Sunshine Coast forest districts. See BC Ministry of Forests 2000d. The ISO pilot was a follow-up to a November 1999 PricewaterhouseCoopers gap analysis of the Small Business Forest Enterprise Program (SBFEP) certification. The PricewaterhouseCoopers report (1999) found that the SBFEP was well positioned for SFM certification but should start with the development of an EMS compatible with the ISO 14001 standard.

22 Innes's study found that certification rarely measures the actual extent of change to the forest resource, whereas the FRPA assesses the on-the-ground effectiveness of forest practices in achieving the province's specific resource values. The investigation also revealed that the FRPA indicators constituted only a subset of the wider range of certification system indicators and might therefore not be recognized globally as evidence that BC's forest are being sustainably managed. See Province of British Columbia 2005, 2.

23 A proposal was presented to the Minister of Forests during this period but never gained traction (interview with David Morel, Ministry of Forests, 14 February 2005).

24 The SBFEP permitted the Ministry of Forests to sell Crown timber competitively to individuals and corporations who were registered in the program. In 2002, the SBFEP was moved into the newly created BC Timber Sales Office. The over 4 million hectares of BC Timber Sales were first ISO 14001–certified and then CSA- and SFI-certified between January 2006 and December 2007.

25 The BC Forest Practices Board is an independent watchdog organization at arm's length from the government that conducts audits to keep both the government and private forest operators publicly accountable for their forest practices.

26 A FPB pilot audit was conducted with Pope and Talbot in 2002, with cooperation from KPMG, to test the alignment of ISO and SFI certification with the FPB audits. In 2003, the board also conducted three audits of companies with ISO certification to see how the compliance and management system audit processes could work together. Overall, the study found that certification is based on different standards and aimed at a different audience. Specific conclusions included the following: (1) Certification audit evidence lacks the same rigour as required by FPB audits (e.g., certification auditors record only exceptions or nonconformity, whereas FPB auditors record all field observations); (2) some efficiencies may be gained by utilizing the certification auditor's review of the licensee's risk management control procedures (e.g., this reduced the sample size and time required in the field on the Pope and Talbot audit), but these potential efficiency gains disappear in an operational area with low inherent risk (e.g., flat versus steep terrain); and (3) ISO certification on its own does not go deep enough into forest landscape issues to be of value to a Forest Practices Board audit. The board also identified the fact that the critical challenge in integrating certification successfully into its compliance audit process was the lack of certification transparency. The third-party certification audit reports needed to include more detail on the audit findings and corrective actions and to be more accessible to the public.

27 Interview with Chris Mosher, FPB, 10 February 2005.

28 Industrial freehold land is land held by individuals or companies with a wood-processing facility. Private woodlot land is land held by individual owners without a wood-processing facility.

29 Of the province's 6.8 million cubic metres of allowable annual softwood cut, 3.3 million cubic metres are from Crown land, 1.9 million cubic metres are industrial freehold, and 1.6 million cubic metres are from private woodlot. See NBDNR 2004, 7, and APEC 2003, 10.

30 Seventeen percent of softwood fibre mill consumption is met by imports (1.5 million cubic metres out of a total of 8.8 million cubic metres consumed per year). See NBDNR 2004, 7.

31 The one exception is the Eel Ground Community Development Centre Inc. The Eel Ground First Nation is a leader in Native forest management. With financial assistance from the federal government, it certified its 2,853-hectare forest (on the north shore of New Brunswick, just outside the city of Miramichi) to the FSC Maritimes regional standard in September 2005.

32 Two sublicensees (North American Forest Products and Groupe Savoie) certified on their own.

33 The province allocates timber volume to sublicensees under several licences. To avoid sublicensees potentially having to certify to several systems, the industry agreed to a uniform certification approach.
34 Interviews with Jennifer Landry, UPM-Kymmene, 18 February 2005; Scott Macdougall, J.D. Irving, 17 February 2005; and Yvon Poitras, New Brunswick Forest Products Association, 14 February 2005.
35 This requirement was adapted in the next version of the CSA standard to better accommodate existing public participation processes.
36 Interview with Jennifer Landry, UPM-Kymmene Miramichi, 18 February 2005. She was also chair of the New Brunswick SFI Implementation Committee (SIC) at this time.
37 Nagaya Forest Restoration Ltd. and Woodlot Stewardship Coop Ltd. FSC-certified a total of 1.385 hectares.
38 In 1994, J.D. Irving had an FSC-certification audit carried out by Scientific Certification Systems (SCS) auditors and was recommended for certification in 1997. The company delayed announcement of its certification until 1999, however.
39 Interview with Doug Mason, New Brunswick Department of Natural Resources (DNR), 9 February 2005.
40 Ibid.
41 The Pan Canadian Woodlot Certification program was under development to complement the CSA, SFI, and FSC programs and to apply to the more than 450,000 private woodlot owners across Canada.
42 The SIC worked to develop, endorse, and set the bar on SFI Best Management Practices, as well as address issues such as the establishment of verification requirements for private land.
43 The supply of fibre from the many woodlots in the province was described by one company as a "spider web"; hence, it was very difficult to track the source.
44 For example, under their forest management agreements, licensees were required to address issues such as deer yards, protected areas, and biodiversity.
45 The mandating of certification on Crown land was implemented as a forest *policy* and not a regulation. Essentially, the policy enforced certification through the government's allocation of Crown timber under the Crown Lands and Forest Act. As explained by the DNR, "it's up to the Minister to allocate the timber harvest to the licensees. Under the policy, if the licensee isn't certified then they won't be allowed to harvest their allocation" (interview with Doug Mason, DNR, 9 February 2005).
46 Industry and government jointly commissioned the Finnish Jaakko Pöyry Consulting (JPC) firm to examine the potential to increase wood supply on New Brunswick Crown land. The JPC report, *New Brunswick Crown Forests: Assessment of Stewardship and Management,* was released in December 2002. In April 2003, an internal process was initiated within the DNR to evaluate the findings of the report. The DNR report, *Staff Review of the Jaakko Pöyry Report,* was released in January 2004.
47 MRNF is the Ministère des Ressources naturelles et de la Faune. Prior to 2006, the ministry also included parks.
48 Groupement Forestier de l'Est du Lac Témiscouata is an enterprise owned by 436 woodlot owners. It serves private woodlots in six municipalities in the Great Lakes and St. Lawrence forest region in the southern part of Quebec.
49 Interview with Bernard Senécal, Domtar, 18 March 2005.
50 The Coulombe Commission was created in October 2003 to examine the economic, environmental, sustainable, social, and regional aspects of Quebec's forests and to analyze long-term technical and scientific calculations and methods for forestry.
51 Interview with Jean Legris and Germain Paré, MRNFP, 6 April 2005.
52 Ibid.
53 The commission found that Quebec's forests were being overharvested and high-graded, that provincial forests information was inadequate, and that the province was lagging behind other jurisdictions in sustainable forest management policy and certification uptake. See MRNFP 2004.
54 Although the Quebec government attended the founding meeting of the FSC in Toronto in 1993, it walked away from the process.

55 Interview with Guy Tremblay, Abitibi-Consolidated, 23 March 2005.

56 Other areas of conflict included the leaving of standing trees and the decommissioning of roads. Certification encouraged these activities and the legislation prohibited them.

57 At this time, Domtar was waiting for a response from the Quebec government in order to proceed with an FSC pilot in Val-d'Or, Quebec.

58 Interview with Jean Legris and Germain Paré, MRNFP, 6 April 2005.

59 Of the total forest area, the government considers 56.8 million hectares to be productive forest.

60 Challenges to Canada's provincial forest economies include the high Canadian dollar, increasing energy costs, and the emergence of competing offshore low-cost forest product producers.

61 Forest owner Peter Schleifenbaum was a member of the FSC Great Lakes/St. Lawrence regional standard technical committee and approached SmartWood in 1997 about conducting an FSC audit on his Haliburton Forest. See "Haliburton Forest – Canada's first certified, sustainable forest," http://www.haliburtonforest.com/forestry_print.html.

62 The Model Forests are a network of community-level research forests across Canada that are supported by the federal Department of Natural Resources.

63 The joint OMNR/FSC International announcement followed from a visit by the Ontario Minister of Natural Resources to FSC headquarters in Oaxaca. Minister Jon Snobelen, FSC International Executive Director Dr. Maharaj Muthoo, and FSC Canada's Director General met for a full day and together drafted the joint press release.

64 Letter from Martin von Mirbach, Director, Sierra Club of Canada, to the FSC Canada Working Group, 29 March 2001.

65 Letter from Arlin Hackman, Vice-President, WWF-Canada, to Jim McCarthy, Executive Director, FSC Canada, 3 April 2001.

66 The Ontario Boreal standard was a pilot regional standard that served as a guidance document for the development of the National Boreal standard.

67 Celia Graham from the OMNR participated on the National Boreal Committee as the Canadian Council of Forest Ministers representative.

68 Interview with Celia Graham, OMNR, 4 February 2005.

69 For example, Abitibi compared the overlap between the province's forest management planning manual and the CSA, and found that twenty-six out of thirty-eight CSA requirements were already addressed.

70 Section 10 of Ontario Regulation 167/95 pertaining to the Crown Forest Sustainability Act was amended on 4 May 2007 and published in the Ontario Gazette on 19 May 2007. Under the amendment, if licensees do not agree with the requirement of certification, they have the opportunity to make representation to the minister before a change is made to their licence agreement.

71 As of 2009, the Quebec government had not yet made any formal announcement to mandate certification.

72 Some scholars (e.g., Elliott 2000, 137; Howlett and Rayner 2001, 49) argue that a form of "triadic network" is emerging in the Canadian forestry sector that includes industry, government, and ENGO members.

73 The film was made by Richard Desjardins and Robert Monderie with the support of Cinéma Libre and the National Film Board of Canada, and released on video in 1999.

74 Spruce budworm is an insect that defoliates northeastern North American forests, targeting balsam fir and white spruce in particular. To control the budworm outbreaks, New Brunswick sprayed the chemical insecticides DDT (1952-70), fenitrothion (1970-95), and Bt (*Bacillus thuringiensis*) (1993 to present). As a result of health and environmental concerns, DDT use was phased out in the 1970s. DDT was banned from registration in Canada in 1985, and fenitrothion was banned in 1998.

Chapter 5: United States

1 Non-industrial private forest is a class of small private lands where the owner does not operate any wood-processing mills. NIPF owners are also referred to as family forest owners.

2 State trust land was granted to the states by the federal government to support public education in the state. While states created before 1850 sold all or nearly all of their federally granted lands, many western states have retained their trust lands. The original thirteen colonies and the next three states contained no federal land and therefore did not receive any federal land grants. See Souder and Fairfax 1995.
3 County and municipal forests constitute 1 percent of US forests and are regulated (particularly in the Northeast) through local ordinances. Western state forest practice acts, however, limit the power of municipal/county governments to pass local ordinances.
4 The federal government's regulatory authority on private forest land is limited to essentially two main areas: protecting habitat for threatened and endangered species (Endangered Species Act), and setting minimal national standards for air and water quality (Clean Air Act; Clean Water Act). In terms of federal public land authority, 60 percent of the federal lands are national forests administered by the US Forest Service (USFS). Historically, the USFS focused on ensuring an annual supply of timber from national forests, but since the early 1990s, its primary role has shifted to helping the states and private landowners achieve voluntary forestry conservation practices. The Department of Interior oversees the Bureau of Land Management (owns commercial forest largely in the western regions of the country) and the National Park Service. The USFS national forests are regulated under the National Forest Management Act and the Bureau of Land Management lands under the federal Forest Land Policy and Management Act.
5 Committees are the most common coordinating mechanism across the US. Commissions are located in the Southeast, councils in the East, and boards in the West. Boards are generally smaller and are accountable to the governor. Councils and commissions are larger and report to the department head. Commissions are slightly larger than boards and are accountable to the legislature rather than the governor. See Kilgore and Ellefson 1992.
6 For example, some states, particularly in the West, place their forestry, parks, and wildlife agencies in separate departments. Great Lakes and Midwestern states house them as divisions in a single natural resources department. Northeastern states typically combine forestry and parks and keep wildlife distinct. Certain states, such as New York and Missouri, combine wildlife and forestry, with parks in a different department. See Ellefson, Hibbard, and Kilgore 2003.
7 For example, in his review of state forest administration in forty-eight states, Koontz (2007) found that twenty-six had statutes with a multi-use mandate while the other twenty-two did not legally specify that forests were to be managed for any particular uses.
8 This figure draws on data from USDA 2002 and Ellefson, Cheng, and Moulton 1995. Comprehensive state forest regulatory programs address water quality, reforestation, timber harvesting, wildfire, insects and disease, rare and endangered species, recreation, and aesthetic values.
9 See Appendix B for a description of the certification programs.
10 The nine FSC regional standards are the Appalachian, Lake States, Mississippi Alluvial Valley, Northeast, Ozark-Ouachita, Pacific Coast, Rocky Mountain, Southeast, and Southwest standards. In 2007, FSC US began developing a national FSC standard to harmonize the nine regional standards. The standard was approved in 2010.
11 In 1993, SCS awarded its first FSC forest management certificate to Collins Pine Company's California division, and in 1994 SmartWood FSC-certified Keweenaw Land Association Ltd. in Michigan.
12 FSC International endorsed the FSC US working group in 1997.
13 The 10 million family forest owners in the US are typically labelled "non-joiners." Most have never harvested their forests and those that do practice active forestry typically consider the costs of certification prohibitive. See Mater 2001. Certification is therefore a huge challenge for this group. FSC and SFI small-landowner options include group certification and master logger certification programs.
14 The study included five case-study national forests: Allegheny National Forest (Pennsylvania), Chequamegon-Nicolet National Forest (Wisconsin), Mt. Hood National Forest (Oregon), three national forests in Florida that are under one forest plan, and the Lakeview Federal

Stewardship Unit on the Fremont-Winema National Forest (Oregon). See Sample et al. 2007 and http://www.fs.fed.us/projects/.

15 For example, the Sierra Club opposed national forest certification, arguing that it would increase harvest, whereas the Nature Conservancy has been in favour, arguing that it would encourage sustainable forest management improvements.

16 Wisconsin has been the lead state in providing direct certification tax incentives to private landowners. For example, under the Wisconsin Managed Forest Law (MFL) group certification program, family forest owners receive tax exemptions and ATFS certification recognition by preparing a DNR-approved long-term forest management plan. As of spring 2007, the MFL program had 2 million ATFS group-certified acres, which made the Wisconsin MFL Tree Farm the largest certified group of private owners in North America. Indiana followed the Wisconsin example with the ATFS group certification of its Classified Forest System program in December 2006.

17 In March 2007, the Washington state Public Lands Commissioner announced that the state would FSC-certify 141,000 acres in the western part of the state. Certification was achieved in May 2008.

18 Interview with Paul Deizman, Forest Management Unit Leader, Division of Forestry, Tennessee Department of Agriculture, 26 October 2006.

19 The states that proceeded with certification following their participation in the Pinchot Institute pilot study were Pennsylvania (1997), Minnesota (2005), Tennessee (2002), Maine (2002), and North Carolina (2001).

20 Interview with Michael Chesnutt, Forest Supervisor, and Hans-Christian Rohr, Management Forester, North Carolina Division of Forest Resources, Department of Environment and Natural Resources, 11 October 2006.

21 Interview with Dan Devlin, Assistant State Forester, Pennsylvania DCNR, 13 October 2006.

22 As outlined in Chapter 3, buyer pressures were mounting during this period. The 1999 announcement by Home Depot that it would give preference to wood from responsibly managed forests by the end of 2002 was followed by similar announcements from large customers such as Lowes, Centex Homes, and Andersen Windows. The Certified Forest Products Council in the US (now called Metafore) was also continuing to work with additional purchasers to develop and adopt forest resource policies that preferentially specified forest products from certified, well-managed forests. As well, AOL Time Warner was leading the market trend towards the adoption of forest sustainability procurement policies among large US paper purchasers such as Hewlett-Packard, Kinko's, Staples, and others, with collective purchases greater than 1.5 tons of paper per year.

23 In July 2003, Maine's governor launched the Maine Forest Certification Initiative, committing the state to increasing the amount of certified forestland in the state to at least 10 million acres by the end of 2007.

24 In November 2003, Time Inc. announced that it would increase its purchases of certified Maine paper from 90,000 tons per year to 120,000 tons by 2005. The company stated that Maine was being praised for its "groundbreaking effort to certify that its forest practices are sustainable." David Refkin, the director of sustainable development for Time Inc., further explained that "Time's strategy is to reward leaders" (interview with David Refkin, 31 May 2007).

25 Interview with Tom Baumann, Assistant to the Director of Forest Management, Division of Forestry, Minnesota DNR, 3 November 2006.

26 Interview with Bob Mather, Director, Bureau of Forest Management, Wisconsin DNR, 25 October 2006.

27 See *Wood Product Industry Trends and Michigan Forests,* a White Paper prepared by the Michigan DNR, 23 June 2005.

28 Comment by Mindy Koch, chief of the Forest, Mineral and Fire Management Division at the Michigan DNR, as reported in Capital News Service, 27 February 2004.

29 The increasing state budget deficits were most clearly seen in year-end balances that plummeted by 70 percent from $37.8 billion in fiscal 2001 to $14.5 billion in fiscal 2003. See NASBO 2002.

30 For example, during the 1990s, Michigan lost 20,000 jobs, $700 million in wages, and over 300 individual businesses or manufacturing facilities from the forest products industry. See Berghorn 2005.
31 The report also recommended that county and private land certification be encouraged and that certification pilot projects be conducted on Minnesota's Chippewa and Superior national forests.
32 Interview with David Todd, State Forester, Division of Forestry, Tennessee Department of Agriculture, 23 October 2006.
33 In October 2004, the Washington Environment Council along with the National Audubon Society, Conservation Northwest, and the Olympic Forest Coalition filed a lawsuit against the state's DNR to overturn the state plan, which included increased harvest levels on state-owned forests (by 30 percent). The groups reached a settlement with the state in March 2006.
34 Based on interviews I conducted in the fall of 2006 with major industrial forest producers and forest industry associations across the US, including Weyerhaeuser, Plum Creek Timber, Stora Enso, MeadWestvaco, Boise Cascade, Canfor, Seven Islands Land Company, Bowater, International Paper, and Domtar, as well as the AF&PA, the Michigan Forest Products Council, and the Minnesota Forest Industries Association (see Appendix A). I asked the industry interviewees the question, "Should state governments certify their state-owned forests? Why?"
35 For example, the Michigan DNR is the largest forestland owner in the state and accounts for 20-25 percent of the state fibre supply. The industry in Michigan was therefore a strong advocate of state forest certification (interview with George Berghorn, Director of Forest Policy, Michigan Forest Products Council, and chair of the Michigan SFI implementation committee, 5 December 2006).
36 For studies on state forest certification implementation costs, see Cubbage et al. 2003; Mater et al. 1999, 2002; Mather 2004.
37 The direct costs included the costs of both audit preparation and the certification inspections.
38 Interview with Paul Deizman, Forest Management Unit Leader, Tennessee Division of Forestry, 26 October 2006.
39 Tennessee certified to the FSC only as its SFI certification audit identified several "non-conformances," largely related to documentation gaps. The state also lacked funds to seek dual certification (interview with David Todd, State Forest System Forester, Tennessee Division of Forestry, 23 October 2006).
40 Interviews with Andrew Arends, Forest Certification Program Coordinator, Minnesota DNR, 24 October 2006; and Tom Baumann, Assistant to the Director, Minnesota DNR, 3 November 2006.
41 Interview with Dennis Nezich, Forest Certification Specialist, Michigan DNR, 16 January 2007.
42 Interviews with US forest companies, spring 2006 to spring 2007 (see Appendix A).
43 For example, among the conditions of the FSC audits conducted in 2001 and 2003, the Washington DNR was required to recalculate its allowable annual cut and reduce harvest levels by extending rotation ages or increasing green retention. The state had just received harvesting approvals under its new Habitat Conservation Plan, however, and was in the midst of its harvest recalculation process.
44 For reports comparing state forest policy and regulations with certification requirements, see Cook and O'Laughlin 2003; Dicus and Delfino 2003; and Fletcher, Adams, and Radosevich 2001.
45 Following its December 2005 FSC pre-assessment audit of Oregon's Klamath County state forestlands, the state's DNR decided not to pursue FSC certification. The reasons included an aversion to a long-term commitment to changing FSC requirements; the FSC requirement to produce a rationale statement on why the state was certifying only one parcel of its forestland; conflicts with several FSC criteria regarding harvest levels, impact analysis, retention, and chemical application; and the FSC requirement for a new management plan every ten years (interviews with Marvin Brown, State Forester, Oregon DNR, 18 October

2006; and David Morman, Program Director, Forest Resources Planning, Oregon DNR, 2 November 2006).

46 Interview with Michael Chesnutt, Forest Supervisor, and Hans-Christian Rohr, Management Forester, North Carolina Division of Forest Resources, Department of Environment and Natural Resources, 11 October 2006.

47 Interview with Cara Boucher, Forest Resource Management Section Manager, Michigan, DNR, 16 January 2007.

48 Interview with Bob Mather, Director, Bureau of Forest Management, Wisconsin DNR, 25 October 2006.

49 Several states expressed concerns about the proposed strengthening of the FSC pesticide policy. Some states were facing ENGO appeals over their state forest certification. For example, ENGOs in Minnesota were appealing the state's FSC certification on the grounds that it did not manage recreational motorized vehicle usage adequately. In 2005, shortly following Michigan's state forest certification, the Sierra Club complained to the FSC that the Michigan DNR was not compliant with FSC principles and criteria. In response, in October 2006 the FSC's Accreditation Services International re-evaluated the Michigan certification and upheld it, arguing that the DNR demonstrated general consistency with the FSC direction.

50 Interview with Dennis Nezich, Forest Certification Specialist, Michigan DNR, 16 January 2007.

51 Interview with Michael Chesnutt, Forest Supervisor, and Hans-Christian Rohr, Management Forester, North Carolina Division of Forest Resources, Department of Environment and Natural Resources, 11 October 2006.

52 Interview with Donald Mansius, Director, Forest Policy and Management, Maine Forest Service, Department of Conservation, 20 October 2006.

Chapter 6: Sweden

1 For studies on forest certification development and governance in Sweden, see Boström 2003; Cashore, Auld, and Newsom 2004; Elliott 2000; Elliott and Schlaepfer 2001; Gulbrandsen 2005a, 2005b; and Klingberg 2003.

2 For example, over the past fifty years there has been a decline in broadleaved forests, a loss of deadwood and an overrepresentation of even-aged uniform forest structure. The government has identified that 200-300 forest-dwelling species are threatened with extinction within 100 years unless appropriate measures are taken. See National Board of Forestry and Swedish Environmental Protection Agency 2003.

3 The right of common access gives citizens freedom to enjoy the countryside but there are limitations. For example, access does not include the use of vehicles. It is also illegal to harm the environment, to cause financial losses to a landowner, or to prevent a landowner from using his or her land.

4 Boreonemoral forest is a transition zone between the coniferous forest of the boreal forest and the mixed coniferous/deciduous forest of the nemoral forest. "Noble" hardwoods, including beech, dominate the nemoral forest.

5 The three major private Swedish forest companies with large productive forestland holdings include SCA (2 million hectares), Holmen (1.035 million hectares), and Bergvik (1.9 million hectares). (Bergvik skog AB was formed in March 2004 and acquired 1.5 million hectares of productive forestland from Stora Enso and 321,000 hectares from Korsnäs.)

6 Four of Sweden's six private landowner associations are coordinated by the Swedish Federation of Family Forest Owners (Skogsägarna LRF). Norra Skogsägarna (North), Skogsägarna Norrskog (North), Skogsägarna Mellanskog (Central), and Södra Skogsägarna (South) are members of LRF. Nätrgaälven Virkesförsäljningsförenin and Västra Värmlands o. Dals Skogsägareförening are independent associations.

7 The market for fibre in Sweden has recently become even more competitive as a result of increasing demands from the bioenergy sector, which is responding to the renewable energy targets of the Swedish government and the EU. The forest industry currently consumes more than half of Sweden's biofuels, and the competitive use of wood fuels for district heating and electricity production is increasing. See Swedish Forest Agency 2007.

8 The regional private independent sawmill associations under the National Federation include Såg i Syd (South), Sågverken I Mellansverge (Central), and SÅGAB and Nedre Norrlands Sågverksförening (North).
9 Sweden imports approximately 15 percent of its industrial wood consumption. Russia is a major supplier, and 80 percent of the imported Russian fibre supply is birch pulpwood.
10 Sweden is an active partner in the EU-FLEGT Action Plan, which provides the basis for trade measures to eliminate illegally logged timber to European markets. For advocacy reports on the issues with respect to Swedish import of illegal fibre, see Lloyd 2000; Lopina, Ptichnikov, and Voropayev 2003; and Taiga Rescue Network 2005.
11 Sustainable Forests is one of Sweden's sixteen environmental objectives under parliament's *Swedish Environmental Objectives – Interim Targets and Action Strategies,* to be achieved by 2020. The "Living Forests" goals are to preserve the national productive capacity of the forest land, maintain the natural function and productivity of forest ecosystems, and maintain viable populations of domestic plant and animal species living in natural conditions.
12 Sveaskog is Sweden's largest public land certification holder (3.4 million hectares). The National Property Board is the country's second-largest public land certification holder, with 1.1 million hectares of FSC-certified forest.
13 In 1995, the Swedish Forest Industries Association launched the Nordic Forest Certification Project in cooperation with representatives from the Finnish and Norwegian forest industries. The project never gained traction, largely because it was actively boycotted by Swedish ENGOs.
14 The FSC working group was chaired by Dr. Lars-Erik Liljelund (senior advisor to the Minister of the Environment) and included six environmental representatives, six economic representatives, and three social representatives.
15 An accelerated FSC consensus was initially reached largely on the basis of "avoided conflict," not necessarily agreement on specific targets. FSC International recognized that the standard lacked resolution and specificity in key areas (protection of old growth, logging of key habitats, and the use of intensive silviculture methods), but it needed a flagship and did not want to discourage the Swedish initiative. It therefore ratified the standard on the understanding that the first round of revisions would address the shortcomings and define the necessary SFM targets.
16 Northern family forest owners opposed the FSC standard mainly because they did not want certification to provide for increased Sámi access rights to reindeer grazing on private forestland. Key issues in the South were largely financial and were related to reduced harvests resulting from key biotope and deciduous forest set-aside requirements and restrictions on insecticide usage. Opponents did not think it was reasonable to expect small forest owners to preserve large key habitats. In addition, a fundamental hurdle was that the FSC chain-of-custody rules required fibre segregation rather than a percentage-based system, thereby making it difficult for the small independent private sawmills to certify their highly fragmented fibre supply chains. Interviews with Tage Klingberg, former chairman of Skogsägarna LRF, 11 September 2007; Folke Stenstrom, former director, PEFC Sweden, 11 September 2007; and Jan-Åke Lunden, chief forester, Skogsägarna LRF, 19 September 2007. See also Elliott 2000, 196; Lindahl 2001, 15.
17 Mellanskog had already established its own regional certification standards for its members (based on ISO 14001), and this, along with other member association environmental standards, served as templates for the development of the Swedish PEFC standard.
18 The PEFC Interim Council included the forest owner associations, the sawmill association, and sections of the Church of Sweden, who participated as observers. ENGOs were invited but chose not to participate.
19 The groups involved in the Stock Dove committee included the Swedish Federation of Family Forest Owners, the Swedish Forest Industries Association, the Swedish Society for Nature Conservation, and the WWF.
20 The three main recommended changes to the PEFC standard based on the Stock Dove process included increasing the nature conservation set-aside requirement from 3 percent

to 5 percent in northern Sweden (implemented on 1 October 2002); a nature value assessment applied to all estates without green management plans or Environmental Consideration Documents; and a key woodlands habitat logging moratorium introduced up to 31 December 2004. See Skogsägarna LRF 2002.

21 For example, AssiDomän commenced certification to the FSC Principles and Criteria in 1996 and by June 1998 had certified all of its 3.3 million hectares of productive forestland. (The eight certified forest divisions of AssiDomän were later consolidated under the certification of Sveaskog.) Stora (prior to the merger with Enso-Gutzeit Oy in 1998) and Korsnäs began their FSC certification processes in 1997 and achieved certification by January 1998. SCA lagged slightly behind the other companies and certified its 1.8 million acres in northern Sweden by the fall of 1998.

22 Interview with Bo Wallin, former head of the Environment Department, SFA, 14 September 2007.

23 Interview with Sune Sohlberg, SEPA, 1 November 2007.

24 Interview with Tage Klingberg, former chairman, Skogsägarna LRF, 11 September 2007.

25 Interview with Jan-Åke Lunden, chief forester, Skogsägarna LRF, 19 September 2007.

26 Interview with Erik Sollander, SFA, 13 September 2007.

27 Interview with Peter Roberntz, former executive director, and Karin Fallman, vice-director, FSC Sweden, 18 September 2007.

28 Interview with Tage Klingberg, former chairman, Skogsägarna LRF, 11 September 2007.

29 According to the FSC working group, nationwide application of the FSC standard would have led to 13.4 percent lower wood production. See Balsiger 1998. See also Eriksson, Sallnas, and Stahl 2007 for a more recent analysis that confirms the 13 percent reduction.

30 Interview with Lena Dahl, TetraPak (formerly with WWF-Sweden), 13 September 2007.

31 See SWEDAC, http://www.swedac.se/sdd/System.nsf/(GUIview)/index_eng.html.

32 Interview with Peter Nohrstedt, Lead Manager EKU, Swedish Environmental Management Council, 23 November 2007.

33 In addition, the Swedish government's shift in regulatory approach was consistent with the deregulatory climate in Europe in the 1990s at the time of the collapse of Eastern European communist systems.

34 Interview with Sune Sohlberg, SEPA, 1 November 2007.

35 A Forest and Environment Declaration provides information about the age and area of the forest stand, as well as details about the forest regeneration activities and environmental data regarding broadleaved forest, nature reserves, protected biotopes, wetlands, archeological sites, and other high conservation value areas within the site. See Wilhelmsson 2006, 53.

36 In terms of accounting, government criteria for voluntary reserves were that they had to be a minimum of five hectares and no subsidy compensation could be provided for the set-aside. Certification reserves could count land that had received compensation, however, and the biodiversity limitations were different. Finally, certification restricted harvest activities within reserve areas, whereas the government policy was less clear.

37 Interview with Erik Sollander, SFA, 13 September 2007.

38 Ibid.

39 Ibid.

40 In this chapter, the term "SFM discourse" refers to the political decision-making process and accepted (institutionalized) beliefs and understandings exchanged by standard forest actors regarding the sustainable management of Sweden's forests.

41 Within the FSC governance structure, economic, social, and environmental stakeholders have equally weighted representation. The PEFC in Sweden also has multi-stakeholder participation, but forest owners and industry have a two-thirds weighted majority vote.

42 See Sandström and Widmark 2007 for a discussion of the role of the FSC in enhancing Sámi consultations.

43 For example, it was suggested that certification had perhaps shut out local groups because they lacked the necessary level of resources and vertical networks to access the national-level forest policy consultation processes. See also Lindahl 2008 for further explanation of

this point. Some interviewees also noted that the efforts of the certification bodies to align their criteria with international conventions and agreements possibly also undermined the degree of local influence.

44 Interview with Erik Sollander, SFA, 13 September 2007.

45 The government's long-term vision included four elements regarding utilization of forest resources to sustain a diversity of values; a high level of fellings and the maintenance of natural ecological processes; a biologically rich forest environment; and social and cultural values. See Swedish Forest Agency 2005.

Conclusion

1 "Virtuous cycle" is a term originating in the economic and business management fields to describe a system of positive feedback. In contrast, a vicious cycle is characterized by negative feedback.

2 In some cases, the marginal cost of going after the poorer-performing actors may exceed the benefits, depending on the size, degree of non-compliance, and access to the group. As well, there will be political factors such as the distribution of costs and benefits that influence the determination of the optimal policy target.

3 It is also important to note that not all leading or compliant forest owners/operators in these regions have *chosen* to certify.

4 The rate of certification adoption is decreasing in the historically leading certified regions while increasing in emerging forest-product-producing regions such as Russia, Brazil, and China. See UNECE/FAO 2008.

5 For example, see Rosoman, Rodrigues, and Jenkins 2008 for a recent Greenpeace International report responding to stakeholder feedback regarding "controversial" FSC certifications.

References

Abusow, K. 2001. Forest certification: Multiple standards advance sustainable forest management. *Wood Design and Building* 18: 42-44.

Albareda, L., Lozano, J.M., and Ysa, T. 2007. Public policies on corporate social responsibility: The role of governments in Europe. *Journal of Business Ethics* 74 (4): 391-407.

Alvarez, M. 2007. *The state of America's forests.* Bethesda, MD: Society of American Foresters.

Anderson, J.W. 1989. *Corporate social responsibility: Guidelines for top management.* New York: Quorum Books.

Ansell, C., and A. Gash. 2008. Collaborative governance in theory and practice. *Journal of Public Administration Research and Theory* 18 (4): 543-71.

Antweiler, W., and K. Harrison. 2007. Canada's voluntary ARET program: Limited success despite industry co-sponsorship. *Journal of Policy Analysis and Management* 26 (4): 755-73.

APEC (Atlantic Provinces Economic Council). 2003. *The New Brunswick forest industry: The potential economic impact of proposals to increase the wood supply, December 2003.* Halifax: APEC.

Auld, G., S. Bernstein, and B. Cashore. 2008. The new corporate social responsibility. *Annual Review of Environment and Resources* 33: 413-35.

Auld, G., L. Gulbrandsen, and C. McDermott. 2008. Certification schemes and the impacts on forests and forestry. *Annual Review of Environment and Resources* 33: 187-211.

Ayres, I., and J. Braithwaite. 1992. *Responsive regulation: Transcending the deregulation debate.* New York: Oxford University Press.

Balsiger, J. 1998. *Swedish forest policy in an international perspective: Summary document.* Messelande 14/1998: Skogsstyrelsen.

Barry, J., and R. Eckersley. 2005. *The state and the global ecological crisis.* Cambridge, MA: MIT Press.

Bartle, I., and P. Vass. 2005. *Self-regulation and the regulatory state: A survey of policy and practice.* Bath, UK: Centre for the Study of Regulated Industries, University of Bath.

BC Ministry of Forests. 1998. Province delivers BC's sustainable forest message in Holland. News release, 13 March. Victoria: Ministry of Forests.

–. 2000a. Advisory council to aid forest management certification. News release, 15 June. Victoria: Ministry of Forests.

–. 2000b. *Annual report Ministry of Forests for fiscal year ended March 31, 2000.* Victoria: Ministry of Forests.

–. 2000c. Premier pledges co-operative strategy on forest marketing. News release, 23 October. Victoria: Ministry of Forests.

–. 2000d. Small business program tests CSA and FSC certification. News release, 31 March. Victoria: Ministry of Forests.

–. 2001. Wilson supports co-operative approach to certification. News release, 23 March. Victoria: Ministry of Forests.

–. 2002a. Changes support timber marketability. News release, 2 May. Victoria: Ministry of Forests.
–. 2002b. *Overview of forest certification, April 2002.* Victoria: Ministry of Forests.
Berghorn, G. H. 2005. *Trends in Michigan's forest products industry 2000-2004.* Lansing: Michigan Forest Products Council.
Bernstein, S. 2005. Legitimacy in global environmental governance. *Journal of International Law and International Relations* 1 (1-2): 139-66.
Bernstein, S., and B. Cashore. 2004. Non-state global governance. In *Hard choices, soft law: Combining trade, environment and social cohesion in global governance,* edited by J. Kirton and M. Trebilcock, 33-63. Aldershot, UK: Ashgate.
–. 2007. Can non-state global governance be legitimate? An analytical framework. *Regulation and Governance* 1 (4): 347-71.
Bichta, C. 2003. *Corporate social responsibility: A role in government policy and regulation?* Bath, UK: University of Bath, School of Management.
Biermann, F., and K. Dingwerth. 2004. Global environmental change and the nation state. *Global Environmental Politics* 4 (1): 1-22.
Bleishwitz, R. 2003. Governance of sustainable development: Towards synergies between corporate and political governance strategies. Working Paper No. 132, Wuppertal Institut fur Klima, Umwelt, Energie, Wuppertal, Germany.
Boström, M. 2003. How state-dependent is a non-state driven rule-making project? The case of forest certification in Sweden. *Journal of Environmental Policy and Planning* 5 (2): 165-80.
–. 2005. Inclusiveness, accountability and responsiveness. Paper presented at the Organizing the World conference, October 2005. Stockholm: Stockholm Centre for Organizational Research.
Bowie, N. 1991. New directions in CSR. *Business Horizons* 34 (4): 56-65.
Braithwaite, J. 2008. *Regulatory capitalism.* Cheltenham, UK: Edward Elgar.
Brinton, M.H., and K.G. Provan. 2000. Governing the hollow state. *Journal of Public Administration Research and Theory* 10 (2): 359-80.
Brinton, M.H., K.G. Provan, and B.A. Else. 1993. What does the "hollow state" look like? In *Public management: The state of the art,* edited by B. Bozeman, 309-22. San Francisco: Jossey-Bass.
Brown, D., and D. Greer. 2001. *Implementing forest certification in British Columbia: Issues and options.* Victoria: Ministry of Forests.
Calder, F., and M. Culverwell. 2005. *Following up the World Summit on Sustainable Development commitments on CSR: Options for action by governments.* London: Chatham House.
Carey, C., and E. Guttenstein. 2008. *Governmental use of voluntary standards: Innovation in sustainability governance.* London: ISEAL Alliance.
Carroll, A.B. 1991. The pyramid of corporate social responsibility: Toward the moral management of organizational stakeholders. *Business Horizons* 34 (4): 39-48.
–. 1999. Corporate social responsibility: Evolution of a definitional construct. *Business and Society* 38 (3): 268-95.
Cashore, B. 2002. Legitimacy and the privatization of environmental governance: How can non-state market-driven (NSMD) governance systems gain rule-making authority. *Governance* 15 (4): 503-29.
Cashore, B., G. Auld, and D. Newsom. 2004. *Governing through markets: Forest certification and the emergence of non-state authority.* New Haven, CT: Yale University Press.
Cashore, B., G. Auld, D. Newsom, and E. Egan. 2009. The emergence of non-state environmental governance in European and North American forest sectors. In *Transatlantic environment and energy politics: Comparative and international perspectives,* edited by M. Schreurs, H. Selin, and S. VanDeveer, 209-30. Surrey, UK: Ashgate.
Cashore, B., S. Bernstein, and C. McDermott. 2007. Conceptualizing forest certification as an environmental cartel: Implications for the institutionalization of globally effective clubs. Unpublished manuscript, 24 June.
Cashore, B., B. Egan, G. Auld, and D. Newsom. 2007. Revising theories of non-state market-driven (NSMD) governance: Lessons from the Finnish forest certification experience. *Global Environmental Politics* 7 (1): 1-44.

Cashore, B., and J. Lawson. 2003. Private policy networks and sustainable forestry policy: Comparing forest certification experiences in the US Northeast and the Canadian Maritimes. *Canadian-American Public Policy* 53 (March): 1-52.

Cashore, B., D. Newsom, F. Gale, and E. Meidinger. 2006. *Confronting sustainability: Forest certification in developing and transitioning societies.* New Haven, CT: Yale School of Forestry and Environmental Studies Publication Series.

CCFM (Canadian Council of Forest Ministers). 1992. *Sustainable forests: A Canadian commitment. National Forest Strategy.* Ottawa: CCFM.

–. 1995. *Defining sustainable forest management: A Canadian approach to criteria and indicators.* Ottawa: Canadian Forest Service, Natural Resources Canada.

CEPI (Confederation of European Paper Industries). 2004. *Forest certification matrix: Finding your way through forest certification schemes.* Brussels: CEPI.

Christian Aid. 2004. *Behind the mask: The real face of corporate social responsibility.* London: Christian Aid.

Christoff, P. 1996. Ecological modernization, ecological modernities. *Environmental Politics* 5 (3): 476-500.

Coady, L. 2002. What I saw of the revolution: Reflections of a corporate environmental manager in the 1990s BC coastal forest industry. In *Bringing business onboard: Sustainable development and the B-school curriculum,* edited by P.N. Nemetz, 539-60. Vancouver: UBC Press.

Coglianese, C., and J. Nash. 2001. *Regulating from the inside: Can environmental management systems achieve policy goals?* Washington, DC: Resources for the Future.

Commonwealth of Australia. 2000. *Critical elements for the assessment of forest management certification schemes.* Canberra: Department of Agriculture, Fisheries and Forestry.

Conroy, M. 2007. *Branded! How the certification revolution is transforming global corporations.* Gabriola Island, BC: New Society Publishers.

Cook, P.S., and J. O'Laughlin. 2003. *Comparison of two forest certification systems and Idaho legal requirements.* Moscow, ID: University of Idaho, College of Natural Resources Policy Analysis Group.

Counsell, S., and K.T. Loraas. 2002. *Trading in credibility: The myth and reality of the Forest Stewardship Council.* London: Rainforest Foundation.

CPET (Central Point of Expertise on Timber). 2006. *Review of forest certification schemes: Results, December 2006.* Oxford: CPET.

CSFCC (Canadian Sustainable Forestry Certification Coalition). 2008. *Canada certification status report, January 31, 2008.* Ottawa: CSFCC.

Cubbage, F.W., and S. Moore. 2008. *Impacts and costs of forest certification: A survey of SFI and FSC in North America.* Paper presented at the 2008 Sustainable Forestry Initiative meeting, 23 September 2008, Minneapolis.

Cubbage, F.W., S. Moore, J. Cox, J. Edeburn, D. Richter, W. Boyette, et al. 2003. Forest certification of state and university lands in North Carolina: A comparison. *Journal of Forestry* 101 (8): 26-31.

Cubbage, F.W., S. Moore, T. Henderson, and M. Araujo. 2008. Costs and benefits of forest certification in the Americas. In *Natural resources: Economics, management and policy,* edited by J.B. Pauling, 155-83. Hauppage, NY: Nova Science Publishers.

Dale, A., and J. Robinson. 1996. *Achieving sustainable development.* Vancouver: UBC Press.

Daly, H. 1990. Toward some operational principles of sustainable development. *Ecological Economics* 2 (1): 1-6.

Dauvergne, P. 1997. *Shadows in the forest – Japan and the politics of timber in Southeast Asia.* Cambridge, MA: MIT Press.

–. 2001. *Loggers and degradation in the Asia Pacific – corporations and environmental management.* Cambridge: Cambridge University Press.

–. 2005. The environmental challenge to loggers in the Asia-Pacific: Corporate practices in informal regimes of governance. In *The business of global environmental governance,* edited by D.L. Levy and P. Newell, 169-96. Cambridge, MA: MIT Press.

den Hertog, J. 2000. *Certification: A view from government.* Presentation to the European Forest Institute Certification Forum, 30 March to 1 April 2000. BC Ministry of Forests.

Dicus, C.A., and K. Delfino. 2003. *A comparison of California forest practice rules and two forest certification systems.* San Luis Obispo, CA: California Polytechnic State University.
Dingwerth, K. 2005. The democratic legitimacy of public-private rule making: What can we learn from the world commission on dams? *Global Governance* 11: 65-83.
Easton, D. 1965. *A framework for political analysis.* Englewood Cliffs, NJ: Prentice-Hall.
Ebeling, J., and M. Yasue. 2009. The effectiveness of market-based conservation in the tropics: Forest certification in Ecuador and Bolivia. *Journal of Environmental Management* 90 (2): 1145-53.
Eckerberg, K. 1990. *Environmental protection in Swedish forestry.* Aldershot, UK: Avebury.
Eckersley, R. 2004. *The green state.* Cambridge, MA: MIT Press.
Eijlander, P. 2005. Possibilities and constraints in the use of self-regulation and co-regulation in legislative policy: Experiences in the Netherlands – lessons to be learned for the EU? *Electronic Journal of Comparative Law* 9 (1).
Elad, C. 2001. Auditing and governance in the forest industry: Between protest and professionalism. *Critical Perspectives on Accounting* 12: 647-71.
Eliadis, P., M.M. Hill, and M. Howlett. 2005. *Designing government: From instruments to governance.* Montreal and Kingston: McGill-Queen's University Press.
Elkington, J. 1998. *Cannibals with forks: The triple bottom line of 21st century business.* Gabriola Island, BC: New Society Publishers.
Ellefson, P.V., A.S. Cheng, and R.J. Moulton. 1995. *Regulation of private forestry practices by state governments.* Station Bulletin 605-1995. St. Paul: Minnesota Agricultural Experiment Station, University of Minnesota.
Ellefson, P.V., C.M. Hibbard, and M.A. Kilgore. 2003. *Federal and state agencies and programs focused on nonfederal forests in the United States: An assessment of intergovernmental roles and responsibilities.* St. Paul: Department of Forest Resources, University of Minnesota.
Ellefson, P.V., M.A. Kilgore, and J.E. Granskog. 2007. Government regulation of forestry practices on private forest land in the United States: An assessment of state government responsibilities and program performance. *Forest Policy and Economics* 9: 620-32.
Elliott, C. 2000. *Forest certification: A policy perspective.* Bogor, Indonesia: Center for International Forestry Research (CIFOR).
Elliott, C., and R. Schlaepfer. 2001. The advocacy coalition framework: Application to the policy process for the development of forest certification in Sweden. *Journal of European Public Policy* 8 (4): 642-61.
Eriksson, L., O. Sallnas, and G. Stahl. 2007. Forest certification and Swedish wood supply. *Forest Policy and Economics* 9 (5): 452-63.
Espach, R. 2006. When is sustainable forestry sustainable? The Forest Stewardship Council in Argentina and Brazil. *Global Environmental Politics* 6 (2): 55-84.
European Commission. 2001. *European governance – A white paper, COM(2001) 428.* Brussels: Commission of the European Communities.
–. 2006. *Implementing the partnership for growth and jobs: Making Europe a pole of excellence on CSR.* Brussels: Communication from the Commission to the European Parliament, the Council, and the European Economic and Social Committee.
–. 2007. *Corporate social responsibility: National public policies in the European Union.* Luxembourg : Employment, Social Affairs and Equal Opportunities, European Commission.
European Economic and Social Committee. 2005. *The current state of co-regulation and self-regulation in the single market.* Brussels: European Economic and Social Committee.
Falkner, R. 2003. Private environmental governance and international relations – exploring the links. *Global Environmental Politics* 3 (2): 72-87.
FERN. 2004. *Footprints in the forest.* Gloucestershire, UK: FERN.
Fletcher, R., P. Adams, and S. Radosevich. 2001. *Comparison of two forest certification systems and Oregon legal requirements.* Report to the Oregon Department of Forestry.
Forest Certification Watch. 2004a. Forest certification in New Brunswick, Part II. *Forest Certification Watch Online.* 26 November. (No longer available online.)
–. 2004b. Quebec: Mandatory certification recommended. *Forest Certification Watch Online.* 17 December. (No longer available online.)

–. 2005. Forest leadership interview: Pierre Corbeil. *Forest Certification Watch Online*. 29 September. (No longer available online.)

Fox, T. 2004. CSR and development: In quest of an agenda. *Development* 47 (3): 29-36.

Fox, T., H. Ward, and B. Howard. 2002. *Public sector roles in strengthening corporate social responsibility: A baseline study*. Washington, DC: World Bank.

FPAC (Forest Products Association of Canada). 2006. *Certification similarities and achievements: Summary comparison of forest certification standards in Canada*. Ottawa: FPAC.

–. 2007. *Forestry certification, FPAC market acceptance customer briefing note*. Ottawa: Forest Products Association of Canada.

FSC (Forest Stewardship Council). 2001. *Clarification on the announcement of Ontario's certification effort*. FSC International press release, 23 April.

FSC-AC (Forest Stewardship Council, Association Civil). 2006. Forest Stewardship Council bylaws. http://www.fsc.org/inst_docs.html.

Gamlin, L. 1988. Sweden's factory forests. *New Scientist* 117 (1597): 41-45.

GAO (US Government Accountability Office). 2005. *Globalization: Numerous federal activities complement US business's global CSR efforts*. Report to Congressional requesters. Washington, DC: GAO.

Gereffi, G., R. Garcia-Johnson, and E.N. Sasser. 2001. The NGO-industrial complex. *Foreign Policy* 125: 56-65.

Gibson, R.B. 1999. *Voluntary initiatives: The new politics of corporate greening*. Peterborough, ON: Broadview Press.

Griffiths, J. 2001. *Proposing an international mutual recognition framework*. Working Group on Mutual Recognition between Credible Sustainable Forest Management Certification Systems and Standards, International Forest Industry Roundtable.

Gulbrandsen, L. 2004. Overlapping public and private governance: Can forest certification fill the gaps in the global forest regime? *Global Environmental Politics* 4 (2): 75-99.

–. 2005a. Explaining different approaches to voluntary standards: A study of forest certification choices in Norway and Sweden. *Journal of Environmental Policy and Planning* 7 (1): 43-59.

–. 2005b. The effectiveness of non-state governance schemes: A comparative study of forest certification in Norway and Sweden. *International Environmental Agreements* 5 (2): 125-49.

–. 2008a. Accountability arrangements in non-state standards organizations: Instrument design and imitation. *Organization* 15 (4): 563-83.

–. 2008b. The role of science in environmental governance: Competing knowledge producers in Swedish and Norwegian forestry. *Global Environmental Politics* 8 (2): 99-122.

Gunningham, N. 2007. Corporate environmental responsibility: Law and the limits of voluntarism. In *The new corporate accountability: Corporate social responsibility and the law*, edited by D. McBarnet, A. Voiculescu, and T. Campbell, 476-500. Cambridge: Cambridge University Press.

Gunningham, N., and P. Grabosky. 1998. *Smart regulation: Designing environmental policy*. Oxford: Clarendon Press.

Gunningham, N., and J. Rees. 1997. Industry self-regulation: An institutional perspective. *Law and Policy* 19 (4): 363-413.

Gunningham, N., and D. Sinclair. 1999. Regulatory pluralism: Designing policy mixes for environmental protection. *Law and Policy* 21 (1): 49-76.

–. 2002a. *Leaders and laggards: Next generation environmental regulation*. Sheffield, UK: Greenleaf Publishing.

–. 2002b. Voluntary approaches to environmental protection: Lessons from the mining and forestry sectors. In *Foreign direct investment and the environment: Lessons from the mining sector*, 157-85. Paris: OCED.

Hajer, M.A. 1995. *The politics of environmental discourse: Ecological modernization and the policy process*. Oxford: Oxford University Press.

Hansen, M., S. Stehman, P. Potapov, T. Loveland, et al. 2008. Humid tropical forest clearing from 2000 to 2005 quantified by using multitemporal and multiresolution remotely sensed data, July 8, 2008. *Proceedings of the National Academy of Sciences* 105 (27): 9439-44.

Harrison, K. 2002a. Ideas and environmental standard-setting: A comparative study of regulation of the pulp and paper industry. *Governance* 15 (1): 65-96.
–. 2002b. Voluntarism and environmental governance. In *Governing the environment: Persistent challenges, uncertain innovations,* edited by E.A. Parson. Toronto: University of Toronto Press.
Haufler, V. 2001. *A public role for the private sector.* Washington, DC: Carnegie Endowment for International Peace.
–. 2003. New forms of governance: Certification regimes as social regulations of the global market. In *Social and political dimensions of forest certification,* edited by E. Meidinger, C. Elliott, and G. Oesten. Remagen-Oberwinter, Germany: Verlag.
Heaton, K. 2001. *Behind the logo: An assessment of the SFI in comparison with the FSC in the USA.* Report prepared for FERN. San Francisco: Natural Resources Defense Council (NRDC).
Held, D., and M. Koenig-Archibugi. 2005. *Global governance and public accountability.* Oxford: Blackwell Publishing.
Hennebel, L., G. Lewkowicz, A. Di Pascale, and B. Frydman. 2007. *Self-regulation and co-regulation of CSR in Europe* (No. ENT/MAP/05/3/3). Brussels: Centre Perelman de Philosophie du Droit Université Libre de Bruxelles.
Heritier, A. 2001. *New modes of governance in Europe: Policy making without legislating?* Bonn: Max Planck Project Group.
Hoffman, A. 2001. *From heresy to dogma: An institutional history of corporate environmentalism.* Palo Alto, CA: Stanford University Press.
Hood, C. 1991. A public management for all seasons? *Public Administration* 69: 3-19.
Howlett, M. 2000. Managing the "hollow state": Procedural policy instruments and modern governance. *Canadian Public Administration* 43 (4): 412-31.
–. 2005. The evolution of de/re/regulation: Spill-overs, perturbations, learning and venue shifting as sources of the re-regulatory imperative. In *De-regulation and its discontents: Rewriting the rules in Asia,* edited by M. Ramesh and M. Howlett. Cheltenham, UK: Edward Elgar.
Howlett, M., and J. Rayner. 2001. The business and government nexus: Principal elements and dynamics of the Canadian forest policy regime. In *Canadian forest policy: Adapting to change,* edited by M. Howlett. Toronto: University of Toronto Press.
Humphreys, D. 1996. *Forest politics: The evolution of international cooperation.* London: Earthscan.
–. 2006. *Logjam: Deforestation and the crisis of global governance.* London: Earthscan.
Hysing, E., and J. Olsson. 2005. Sustainability through good advice? Assessing the governance of Swedish forest biodiversity. *Environmental Politics* 14 (4): 510-26.
Ingemarson, F. 2004. Small-scale forestry in Sweden – owner's objectives, silviculture practices and management plans. PhD dissertation, Swedish University of Agricultural Sciences (SLU), Uppsala.
Jenkins, R. 2001. *Corporate codes of conduct: Self-regulation in a global economy.* Geneva: United Nations Research Institute for Social Development.
Jessop, B. 2002. Governance and meta-governance: On reflexivity, requisite variety and requisite irony. Department of Sociology, Lancaster University, http://www.lancs.ac.uk/fass/sociology/papers/jessop-governance-and-metagovernance.pdf.
Jordan, A., R.K.W. Wurzel, and A.R. Zito. 2003. *"New" instruments of environmental governance? National experiences and prospects.* London: Frank Cass Publishers.
–. 2005. The rise of "new" policy instruments in comparative perspective: Has governance eclipsed government? *Political Studies* 53: 477-96.
Kagan, R., N. Gunningham, and D. Thornton. 2003. Explaining corporate environmental performance: How does regulation matter? *Law and Society Review* 37: 51-89.
Kahn, S., and E. Minnich. 2005. *The fox in the henhouse: How privatization threatens democracy.* San Francisco: Berrett-Koehler Publishers.
Karkkainen, B.C. 2004. Post-sovereign environmental governance. *Global Environmental Politics* 4 (1): 72-96.
Kaufmann, D., A. Kraay, and M. Mastruzzi. 2007. *Governance indicators: Where are we? Where should we be going?* World Bank Policy Research Group Working Paper. Washington, DC: World Bank.

Keck, M., and K. Sikkink. 1998. *Activists beyond borders: Advocacy networks in international politics.* Ithaca, NY: Cornell University Press.

Keohane, R.O. 2003. Global governance and democratic accountability. In *Taming globalization: Frontiers of governance,* edited by D. Held and M. Koenig-Archibugi, 130-59. Cambridge: Polity Press.

Kickert, W.J.M., E.-H. Klijn, and J.F.M. Koppenjan. 1997. *Managing complex networks: Strategies for the public sector.* London: Sage Publications.

Kilgore, M.A., and P.V. Ellefson. 1992. *Co-ordination of forest resource policies and programs: Evaluation of administrative mechanisms used by state governments.* Report no. SB-05876. St. Paul: Communication and Educational Technology Services, University of Minnesota Extension.

Klingberg, T. 2003. Certification of forestry: A small forester perspective. *Small-Scale Forest Economics, Management and Policy* 2 (3): 409-21.

Knill, C., and D. Lehmkuhl. 2002. Private actors and the state: Internationalization and changing patterns of governance. *Governance* 15 (1): 41-63.

Kobrin, S. 1998. Back to the future: Neo-medievalism and the postmodern digital world economy. *Journal of International Affairs* 51 (2): 361-87.

Koenig-Archibugi, M. 2005. Transnational corporations and public accountability. In *Global governance and public accountability,* edited by D. Held and M. Koenig-Archibugi, 110-35. Oxford: Blackwell Publishing.

Koleva, M. 2006. *Forest certification – do governments have a role? Proceedings and summary of discussion.* Geneva: United Nations Economic Commission for Europe/Food and Agriculture Organization of the United Nations.

Kooiman, J. 1993. *Governance: New government-society interactions.* London: Sage Publications.

–. 2003. *Governing as governance.* London: Sage Publications.

Koontz, T.M. 2007. Federal and state public forest administration in the new millennium: Revisiting Herbert Kaufman's *The Forest Ranger. Public Administration Review* 67 (1): 152-64.

Koppenjan, J.F.M., and E.-H. Klijn. 2004. *Managing uncertainties in networks.* London: Routledge.

Leipziger, D. 2003. *The corporate responsibility code book.* Sheffield, UK: Greenleaf Publishing.

Levi-Faur, D. 2005. The global diffusion of regulatory capitalism. *Annals of the American Academy of Political and Social Science* 598 (1): 12-32.

Lindahl, K. 2001. *Behind the logo: The development, standards and procedures of the FSC and PEFC scheme in Sweden.* A report prepared for FERN, UK, April. Jokkmokk, Sweden.

–. 2008. Frame analysis, place perceptions and the politics of natural resource management. PhD dissertation, Faculty of Natural Resources and Agricultural Sciences, Swedish University of Agricultural Sciences, Uppsala.

Lipschutz, R.D., and C. Fogel. 2002. Regulation for the rest of us? Global civil society and the privatization of transnational regulation. In *The emergence of private authority in the international system,* edited by R.B. Hall and T.J. Biersteker, 115-40. Cambridge: Cambridge University Press.

Lipschutz, R.D., and J.K. Rowe. 2005. Paper or plastic? The privatization of global forestry regulation. In *Globalization, governmentality and global politics: Regulation for the rest of us?,* edited by R.D. Lipschutz and J.K. Rowe, 106-29. New York: Routledge.

Lister, M., and D. Marsh. 2006. Conclusion. In *The state: Theories and issues,* edited by C. Hay, M. Lister, and D. Marsh, 248-61. New York: Palgrave Macmillan.

Lloyd, S. 2000. *Towards responsible Swedish timber trade? A survey of actors and origin of timber from Russia and the Baltic States.* A report prepared by Sarah Lloyd for the Taiga Rescue Network and WWF-Sweden. Jokkmokk, Sweden: Taiga Rescue Network Publications.

Lopina, O., A. Ptichnikov, and A. Voropayev. 2003. *Illegal logging in Northwestern Russia and export of Russian forest products to Sweden.* Moscow: Worldwide Fund for Nature Russian Programme Office (WWF-Russia).

Lyon, T.P., and J.W. Maxwell. 2002. Voluntary approaches to environmental regulation: A survey. In *Economic institutions and environmental policy,* edited by M. Franzini and A. Nicita, 142-74. Aldershot, UK: Ashgate Publishing.

–. 2004. *Corporate environmentalism and public policy.* Cambridge: Cambridge University Press.

Maine. 2002. *Forest certification in Maine: Report of the Speaker's Advisory Council on forest certification.* Augusta: Office of the Speaker of the House.

Mater, C.M. 2001. Non-joiner NIPFs: What drives their decisions to fragment and/or convert their forestland. A presentation to the Pinchot Institute for Conservation, funded by the Wood Education Research Center, March, Washington, DC.

Mater, C.M., V.A. Sample, J. Grace, and G. Rose. 1999. Third-party, performance-based certification: What public forestland managers should know. *Journal of Forestry* 97 (2): 6-12.

Mater, C.M., W. Price, and V.A. Sample. 2002. *Certification assessments on public and university lands: A field-based comparative evaluation of the FSC and the SFI programs.* Washington, DC: Pinchot Institute for Conservation.

May, E., and R.E.L. Rogers. 1982. *Budworm battles: The fight to stop the aerial spraying of the forests of eastern Canada.* Halifax: Four East.

May, P. 2005. Regulation and compliance motivations: Examining different approaches. *Public Administration Review* 65 (1): 31-44.

Mayntz, R. 2003. *From government to governance: Political steering in modern societies.* Paper presented at the Institute for Ecological Economy Research (IOEW) Summer Academy on integrated product policy (IPP), 7-11 September, Würzburg, Germany.

McCrudden, C. 2007. Corporate social responsibility and public procurement. In *The new corporate accountability: Corporate social responsibility and the law,* edited by D. McBarnet, A. Voiculescu, and T. Campbell, 93-118. Cambridge: Cambridge University Press.

McDermott, C., E. Noah, and B. Cashore. 2008. Differences that "matter"? A framework for comparing environmental certification standards and government policies. *Journal of Environmental Policy and Planning* 10 (1): 47-70.

McKague, K., and W. Cragg. 2007. *Compendium of ethics codes and instruments of corporate responsibility.* Toronto: Schulich School of Business, York University.

Meadowcroft, J. 2000. Sustainable development: A new(ish) idea for a new century. *Political Studies* 48: 370-87.

Meidinger, E. 1997. Look who's making the rules: International environmental standard-setting by non-governmental organizations. *Human Ecology Review* 4 (1): 52-54.

–. 1999. Private environmental regulation, human rights and community. *Buffalo Environmental Law Journal* 7: 123-237.

–. 2003a. Forest certification as a global civil society regulatory institution. In *Social and political dimensions of forest certification,* edited by E. Meidinger, C. Elliott, and G. Oesten, 265-90. Remagen-Oberwinter: Verlag.

–. 2003b. Forest certification as environmental law-making by global civil society. In *Social and political dimensions of forest certification,* edited by E. Meidinger, C. Elliott, and G. Oesten, 293-330. Remagen-Oberwinter: Verlag.

–. 2008. Competitive supragovernmental regulation: How could it be democratic? *Chicago Journal of International Law* 8 (2): 513-34.

Meridian Institute. 2001. *Comparative analysis of the FSC and SFI certification programs.* Washington, DC: Meridian Institute.

Metafore. 2004. *Matching business values with forest certification systems.* Portland, OR: Metafore.

Miles, E.L., A. Underdal, S. Andersen, S.J. Wettestad, J.B. Skjaerseth, and E.M. Carlin. 2002. *Environmental regime effectiveness: Confronting theory with evidence.* Cambridge, MA: MIT Press.

Millennium Ecosystem Assessment. 2005. *Ecosystems and human well-being: Current state and trends – findings of the Condition and Trends Working Group.* Washington, DC: World Resources Institute; Island Press.

Minnesota DNR (Department of Natural Resources). 2003. *Governor's Advisory Task Force report on the competitiveness of Minnesota's primary forest products industry,* July 2003. St. Paul: Minnesota DNR.

Mitchell, R.B. 2002. A quantitative approach to evaluating international environmental regimes. *Global Environmental Politics* 2 (4): 58-83.
Mol, A.P.J., and D.A. Sonnenfeld. 2000. *Ecological modernization around the world: Perspectives and critical debate*. Portland, OR: Routledge.
Moon, J. 2002a. Government as a driver of corporate social responsibility: The UK in comparative perspective. International Centre for Corporate Social Responsibility (ICCSR), Research Paper Series 20-2004, University of Nottingham.
–. 2002b. The social responsibility of business and new governance. *Government and Opposition* 37 (3): 385-408.
Morgenstern, R.D., and W.A. Pizer. 2007. *Reality check: The nature and performance of voluntary environmental programs in the US, Europe and Japan*. Washington, DC: Resources for the Future.
MRNF (Ministère des Ressources naturelles et de la Faune). 2008. *Forests: Building a Future for Quebec*. Quebec City: MRNF (Ministry of Natural Resources and Wildlife).
MRNFP (Ministère des Ressources naturelles, de la Faune et des Parcs). 2003. *La certification forestière*. Quebec City: MRNFP (Ministry of Natural Resources, Wildlife and Parks).
–. 2004. *Commission for the study of public forest management in Quebec, final report summary, December 2004*. Quebec City: MRNFP (Ministry of Natural Resources, Wildlife and Parks).
NASBO (National Association of State Budget Officers). 2002. *Fiscal survey of the states, November 2002*. Washington, DC: NASBO.
National Board of Forestry and Swedish Environmental Protection Agency. 2003. *Protecting the forests of Sweden*. Stockholm: National Board of Forestry and the Swedish Environmental Protection Agency.
Natural Resources Canada. 2004a. *The state of Canada's forests 2003-2004*. Ottawa: Natural Resources Canada.
–. 2004b. *Corporate social responsibility: Lessons learned*. Ottawa: Natural Resources Canada.
–. 2007. *The state of Canada's forests, annual report 2007*. Ottawa: Canadian Forest Service.
NBDNR (New Brunswick Department of Natural Resources). 2004. *DNR staff review of Jaakko Pöyry report, January 2004*. Fredericton: NBDNR.
Nelson, J. 2006. *The public role of private enterprise: Risks, opportunities and new modes of engagement*. Cambridge, MA: John F. Kennedy School of Government, Harvard University.
Newell, P.J. 2001. New environmental architectures and the search for effectiveness. *Global Environmental Politics* 1 (1): 35-44.
–. 2002. From responsibility to citizenship: Corporate accountability for development. *IDS Bulletin* 33 (2): 91-100.
Newsom, D., V. Bahn, and B. Cashore. 2006. Does forest certification matter? An analysis of operation-level changes required during the SmartWood certification process in the United States. *Forest Policy and Economics* 9 (3): 197-208.
Nussbaum, R., and M. Simula. 2004. *Forest certification: A review of impacts and assessment frameworks*. New Haven, CT: Forests Dialogue, Yale School of Forestry and Environmental Studies.
OECD (Organisation for Economic Co-operation and Development). 1999. *Codes of corporate conduct: An inventory*. Paris: OECD.
–. 2003. *Voluntary approaches for environmental policy: Effectiveness, efficiency and usage in policy mixes*. Paris: OECD.
Ohmae, K. 1995. *The end of the nation state*. New York: Free Press.
OMNR (Ontario Ministry of Natural Resources). 1999. *1999 Ontario forest accord: A foundation for progress*. Toronto: OMNR.
–. 2001. Ontario first in the world to receive environmental forest certification. Press release, 23 March 2001. Toronto: OMNR.
–. 2002a. Ontario forest certification memorandum of understanding will assist the province's forest industry. Fact sheet, 7 November 2002. Toronto: OMNR.
–. 2002b. Ontario reaffirms its commitment to sustainable forestry. Press release, 31 January 2002. Toronto: OMNR.
–. 2006a. Collaborative action plan – Ontario Ministry of Natural Resources and the Forest Stewardship Council–Canada, February 2006. Toronto: OMNR and FSC-Canada.

–. 2006b. *Forest process streamlining task force report – implementation plan, October 2006.* Toronto: OMNR.

Osborne, D., and T. Gaebler. 1993. *Reinventing government: How entrepreneurial spirit is transforming the public sector.* Reading: Addison-Wesley.

Ozinga, S. 2004. Time to measure the impacts of certification on sustainable forest management. *Unasylva* 55 (219): 33-38.

Palzer, C., and A. Scheuer. 2004. Self-regulation, co-regulation, public regulation. In *Yearbook 2004,* edited by UNESCO Clearinghouse. Paris: UNESCO.

Parker, C. 2007. Meta-regulation: Legal accountability for corporate social responsibility. In *The new corporate accountability: Corporate social responsibility and the law,* edited by D. McBarnet, A. Voiculescu, and T. Campbell, 207-37. Cambridge: Cambridge University Press.

Pattberg, P. 2006. *The transformation of global business regulation.* Global Governance Working Paper Series, No.18. Amsterdam: Global Governance Project.

–. 2007. *Private institutions and global governance.* Cheltenham, UK: Edward Elgar.

Peters, G.B. 1994. Managing the hollow state. *International Journal of Public Administration* 17 (3-4): 739-56.

–. 2006. *The meta-governance of policy networks: Steering at a distance, but still steering.* Paper presented at the Democratic Network Governance in Europe conference, 2-3 November, Roskilde, Denmark.

Peters, G.B., and J. Pierre. 1998. Governance without government? Rethinking public administration. *Journal of Public Administration Research and Theory* 8 (2): 223-43.

Phidd, R.W., and G.B. Doern. 1983. *Canadian public policy: Ideas, structures, processes.* Toronto: Methuen.

Pinchot Institute for Conservation. 2006. *Oregon forestlands and the Programme for the Endorsement of Forest Certification (PEFC): An assessment of the process and basis for eligibility.* Washington, DC: Pinchot Institute for Conservation.

Porter, M., and C. van der Linde. 1995. Green and competitive: Ending the stalemate. *Harvard Business Review* 73 (5): 120-34.

Potoski, M., and A. Prakash. 2002. Protecting the environment: Voluntary regulations in environmental governance. *Policy Currents* 11 (4): 9-14.

Prakash, A. 2001. Why do firms adopt beyond-compliance environmental policies? *Business Strategy and the Environment* 10: 286-99.

Province of British Columbia. 2005. *Forest certification and the FRPA resource evaluation program, Extension Note #5, January 2005.* Victoria: Province of British Columbia.

Quebec Forest Industry Council. 2004. The QFIC recommends making certification of forestry practices compulsory. Media release, 15 April 2004. Quebec City: Quebec Forest Industry Council.

Rametsteiner, E. 2000. *The role of governments in SFM certification.* Vienna: Institute of Forest Sector Policy and Economics, University of Agricultural Sciences.

–. 2002. The role of governments in forest certification – a normative analysis based on new institutional economics theories. *Forest Policy and Economics* 4: 163-73.

Rametsteiner, E., and M. Simula. 2003. Forest certification – an instrument to promote sustainable forest management. *Journal of Environmental Management* 67 (1): 87-98.

Rhodes, R.A.W. 1996. The new governance: Governing without government. *Political Studies* 44: 652-67.

Rickenbach, M., and C. Overdevest. 2006. More than markets: Assessing Forest Stewardship Council (FSC) as a policy tool. *Journal of Forestry* 104 (3): 143-47.

Roberts, G. 2007. Forest certification faces hard questions. *Ends Report* 395 (December): 38-41.

Rosenau, J., and E.O. Czempiel. 1992. *Governance without government: Order and change in world politics.* Cambridge: Cambridge University Press.

Rosoman, G., J. Rodrigues, and A. Jenkins. 2008. *Holding the line: Recommendations and progress to date on certification body and FSC performance following a critical analysis of a range of "controversial" certificates.* Amsterdam: Greenpeace International.

Ruggie, J.G. 2004. *CSR and global governance: Drivers and trends.* Cambridge, MA: John F. Kennedy School of Government, Harvard University.

Sabatier, P. 1986. Top-down and bottom-up approaches to implementation research. *Journal of Public Policy* 6: 21-48.

–. 1999. *Theories of the policy process.* Boulder, CO: Westview Press.

Salamon, L.M. 2002. *The tools of government: A guide to the new governance.* New York: Oxford University Press.

Sample, V.A., W. Price, J.S. Donnay, and C.M. Mater. 2007. *National forest certification study: An evaluation of the applicability of FSC and SFI standards on five national forests.* Washington, DC: Pinchot Institute for Conservation.

Sanberg, L.A., and P. Clancy. 2002. Politics, science, and the spruce budworm in New Brunswick and Nova Scotia. *Journal of Canadian Studies* 37 (2): 1-22.

Sandström, C., and C. Widmark. 2007. Stakeholders' perceptions of consultations as tools for co-management: A case study of the forestry and reindeer herding sectors in northern Sweden. *Forest Policy and Economics* 10 (1-2): 23-35.

Savcor. 2005. *Effectiveness and efficiency of FSC and PEFC forest certification on pilot areas in Nordic countries.* A report for the Federation of Nordic Forest Owners' Organizations. Helsinki: Savcor Indufor Oy.

Schulz, W., and T. Held. 2004. *Regulated self-regulation as a form of modern government.* Eastleigh, UK: John Libbey Publishing for the University of Luton Press.

Segura, G. 2004. *Forest certification and governments: The real and potential influence on regulatory frameworks and forest policies.* Washington, DC: Forest Trends.

Senden, L. 2005. Soft law, self-regulation and co-regulation in European law: Where do they meet? *Electronic Journal of Comparative Law* 9 (1).

Senes Consultants. 2006. *Review of the independent forest audit, final report, December 2006.* Report prepared by Senes Consultants Ltd. for the Ontario Ministry of Natural Resources.

Seymour, R.S. 2006. *Certification and silviculture – Has anything really changed?* Presentation to the New England Society of American Foresters' (SAF) annual meeting, Nashua, NH, 5 April 2006.

Simula, M. 2006. *Public procurement policies for forest products and their impacts.* Geneva: Forest Products and Economics Division, Food and Agriculture Organization of the United Nations.

Skogsägarna LRF. 2002. Implementation of the agreed bridging document Stock Dove in the PEFC forest certification scheme in Sweden. Press release, 7 February 2002.

Sørensen, E., and J. Torfing. 2007. *Theories of democratic network governance.* New York: Palgrave Macmillan.

Sørensen, G. 2004. *The transformation of the state: Beyond the myth of retreat.* Basingstoke, UK: Palgrave Macmillan.

Souder, J., and S. Fairfax. 1995. *The state trust lands: History, management and sustainable use.* Lawrence: University Press of Kansas.

Spaargaren, G., and A.P.J. Mol. 1992. Sociology, environment and modernity: Ecological modernization as a theory of social change. *Society and Natural Resources* 5 (4): 323-44.

Sprinz, D., and C. Helm. 1999. The effect of global environmental regimes: A measurement concept. *International Political Science Review* 20 (4): 359-69.

Stoker, G. 1998. Governance as theory: Five propositions. *International Social Science Journal* 50 (155): 17-28.

Strange, S. 1996. *The retreat of the state: The diffusion of power in the world economy.* Cambridge: Cambridge University Press.

Suchman, M.C. 1995. Managing legitimacy: Strategic and institutional approaches. *Academy of Management Review* 20 (3): 571-610.

SustainAbility. 2004. *Gearing up: From corporate responsibility to good governance and scalable solutions.* London: SustainAbility.

Swedish Environmental Protection Agency. 2007. *Sweden's environmental objectives de facto 2007 – a progress report from the Swedish Environmental Objectives Council.* Bromma: Swedish Environmental Protection Agency.

Swedish Forest Agency. 2005. *Quantitative targets of Swedish forest policy.* Jönköping: Swedish Forest Agency.

–. 2007. *The statistical yearbook of forestry.* Jönköping: Swedish Forest Agency.

Swift, T., and S. Zadek. 2002. *Corporate responsibility and the competitive advantage of nations.* London: AccountAbility.

Taiga Rescue Network. 2005. *Sweden: Forest industry giant with big timber footprints in the Baltic region.* Helsinki: Taiga Rescue Network.

Tikina, A. 2008. Forest certification: Are we there yet? *CAB Reviews: Perspectives in Agriculture, Veterinary Science, Nutrition and Natural Resources* 3 (18): 1-8.

Tikina, A., R. Kozak, G. Bull, and B. Larson. 2009. Perceptions of change in the US Pacific Northwest forest practices on certified and non-certified holdings. *Western Journal of Applied Forestry* 24 (4): 187-92.

Tollefson, C., F. Gale, and D. Haley. 2008. *Setting the standard: Certification, governance, and the Forest Stewardship Council.* Vancouver: UBC Press.

UNECE/FAO (United Nations Economic Commission for Europe/Food and Agriculture Organization of the United Nations). 2002. *Forest certification update for the UNECE region, summer 2002.* Prepared by J. Raunetsalo, H. Juslin, E. Hansen, and K. Forsyth. Geneva: UNECE/FAO.

–. 2007. *Forest products annual market review 2006-2007, ECE/TIM/SP/22.* Geneva: UNECE/FAO.

–. 2008. *Forest products annual market review 2007-2008.* Geneva: UNECE/FAO.

USDA (US Department of Agriculture). 2002. *Forest resources of the United States, 2002 (a technical document supporting the USDA FS 2005 update of the RPA assessment).* St. Paul, MN: North Central Research Station, Forest Service, USDA.

Utting, P. 2003. Promoting development through CSR – prospects and limitations. *Global Future* 3rd Quarter: 11-13.

–. 2005. *Rethinking business regulation: From self-regulation to social control.* Geneva: United Nations Research Institute for Social Development.

Vogel, D. 2005. *The market for virtue: The potential and limits of corporate social responsibility.* Washington, DC: Brookings Institution.

–. 2006. *The private regulation of global corporate conduct.* Paper presented at the annual meeting of the American Political Science Association, 31 August, Philadelphia, Pennsylvania.

Volpe, J. 2000. *Forest management practices in Canada as an international trade issue.* Report of the Standing Committee on Natural Resources and Government Operations. Ottawa: House of Commons.

Ward, H. 2004. *Public sector roles in strengthening corporate social responsibility: Taking stock.* Washington, DC: World Bank.

Webb, K. 1999. Voluntary initiatives and the law. In *Voluntary initiatives: The new politics of corporate greening,* edited by R. Gibson, 32-50. Peterborough, ON: Broadview Press.

–. 2005. *Sustainable governance in the twenty-first century: Moving beyond instrument choice.* Montreal and Kingston: McGill-Queen's University Press.

Webb, K., and A. Morrison. 2005. The law and voluntary codes: Examining their "tangled web." In *Voluntary codes: Private governance, the public interest and innovation,* edited by K. Webb, 97-174. Ottawa: Carleton Research Unit for Innovation, Science and Environment, Carleton University.

Wettestad, J. 2001. Designing effective environmental regimes: The conditional keys. *Global Governance* 7 (3): 317-42.

Wilhelmsson, E. 2006. Forest management planning for private forest owners in Sweden. *Finnish Forest Research Institute Working Papers* 38: 52-60.

Wolf, K.D. 2006. Private actors and the legitimacy of governance beyond the state. In *Governance and democracy: Comparing national, European and international experiences,* edited by A. Benz and Y. Papadopoulos, 200-27. New York: Routledge.

World Bank. 2003. *Company codes of conduct and international standards: An analytical comparison, October 2003 and January 2004.* Washington, DC: World Bank; International Finance Corporation.

World Business Council for Sustainable Development. 2001. *The business case for sustainable development.* Geneva: World Business Council for Sustainable Development.

World Commission on Environment and Development. 1987. *Our common future*. New York: Oxford University Press.

World Economic Forum. 2008. *Partnering to strengthen public governance*. Geneva: World Economic Forum.

WWF (Worldwide Fund for Nature). 2001. *PEFC – An analysis, a WWF discussion paper*. Zurich: WWF.

–. 2005. *The effects of FSC-certification in Sweden: An analysis of corrective action requests*. A study conducted by Peter Hirschberger, WWF-Austria for the WWF European Forest Programme, February 2005.

WWF/World Bank. 2006. *Forest certification assessment guide (FCAG), July 2006*. Washington, DC: WWF; World Bank Global Forest Alliance.

Young, O.R. 1999. *The effectiveness of international environmental regimes: Causal connections and behavioral mechanisms*. Cambridge, MA: MIT Press.

–. 2001. Inferences and indices: Evaluating the effectiveness of international environmental regimes. *Global Environmental Politics* 1 (1): 99-121.

Zadek, S. 2001. *The civil corporation: The new economy of corporate citizenship*. London: Earthscan.

Zappala, G. 2003. *Corporate citizenship and the role of government: The public policy case*. Research Paper No. 4 2003-04. Canberra: Department of Parliamentary Library.

Index

References to tables and figures are indicated by "t" and "f" following the page number.

Printed and bound in Canada by Friesens

Set in Stone by Artegraphica Design Co. Ltd.

Copy editor: Frank Chow

Proofreader: Kirsten Craven

Indexer: Judith Anderson

Cartographer: Eric Leinberger